3rd International Conference on Mechanical and Aeronautical Engineering (ICMAE 2017)

AA006752

IOP Conference Series: Materials Science and Engineering
Volume 326

Dubai, United Arab Emirates
13 – 16 December 2017

ISBN: 978-1-5108-5953-1
ISSN: 1757-8981

Printed from e-media with permission by:

Curran Associates, Inc.
57 Morehouse Lane
Red Hook, NY 12571

Some format issues inherent in the e-media version may also appear in this print version.

Printed by Curran Associates, Inc. (2018)

For permission requests, please contact the Institute of Physics
at the address below.

Institute of Physics
Dirac House, Temple Back
Bristol BS1 6BE UK

Phone: 44 1 17 929 7481
Fax: 44 1 17 920 0979

techtracking@iop.org

Additional copies of this publication are available from:

Curran Associates, Inc.
57 Morehouse Lane
Red Hook, NY 12571 USA
Phone: 845-758-0400
Fax: 845-758-2633
Email: curran@proceedings.com
Web: www.proceedings.com

Table of contents

Volume 326

3rd International Conference on Mechanical and Aeronautical Engineering (ICMAE 2017)

13–16 December 2017, Dubai, UAE

Accepted papers received: 22 February 2018
Published online: 12 March 2018

Preface

Papers

Mechanical

Aeronautical Engineering

IOP Conf. Series: Materials Science and Engineering **326** (2018) 011001 doi:10.1088/1757-899X/326/1/011001

PREFACE

ICMAE 2017 was held in Dubai, UAE, December 13-16, 2017. ICMAE 2017 was organized by Hong Kong Society of Mechanical Engineers.. The conference provides a useful and wide platform both for display the latest research and for exchange of research results and thoughts in Mechanical and Aeronautical Engineering. The participants of the conference were from almost every part of the world, with background of either academia or industry, even well-known enterprise. The success and prosperity of the conference is reflected high level of the papers received.

The proceedings are a compilation of the accepted papers and represent an interesting outcome of the conference. This book cover 2 chapters: Mechanical and Aeronautical Engineering.

We would like to acknowledge all of those who supported ICMAE 2017. Each individual and institutional help were very important for the success of this conference. Especially we would like to thank the organizing committee for their valuable advices in the organization and helpful peer review of the papers.

We sincerely hope that ICMAE 2017 will be a forum for excellent discussions that will put forward new ideas and promote collaborative researches. We are sure that the proceedings will serve as an important research source of references and the knowledge, which will lead to not only scientific and engineering progress but also other new products and processes.

Prof. Y. Lawrence Yao, Columbia University, USA
Prof. Du Ruxu, The Chinese University of Hong Kong, Hong Kong

IOP Conf. Series: Materials Science and Engineering **326** (2018) 011002 doi:10.1088/1757-899X/326/1/011002

COMMITTEES

Conference Chairman

Prof. Y. Lawrence Yao, Columbia University, USA

Conference Committee Co-Chairs

Prof. Dan Zhang,York University, Canada

Prof. Ruxu Du,The Chinese University of Hong Kong, Hong Kong

Prof. Noha Mohamed Hassan, American University of Sharjah, UAE

Program Committee Co-Chairs

Prof. Matthew Parkinson, Pennsylvania State University, USA

Prof. Ian McAndrew, Embry Riddle Aeronautical University, UK

Advisory Committee

Prof. Peter X. Liu, Carleton University, Canada

Prof. Kai Cheng, Brunel University, UK

International Technical Committee

Prof. Craig Law, University of the Witwatersrand, South Africa

Dr. Anthony R. McAndrew, Cranfield University, UK

Dr. Max BAKAEV, Novosibirsk State Technical University , Russia

Dr. Andrew Carruthers, University of Bradford, UK

Dr. Elena Navarro, Embry Riddle Aeronautical University, USA

Dr. Emily Faulconer, Embry-Riddle Aeronautical University, UK

Dr. Eltayeb Eljack,Aeronautical Engineering Department at King AbdulAziz University, Saudi Arabia

Dr. Redouane Zitoune, Clément Ader Institute of Toulouse University, Toulouse, France

IOP Conf. Series: Materials Science and Engineering **326** (2018) 011002 doi:10.1088/1757-899X/326/1/011002

Peer review statement

All papers published in this volume of *IOP Conference Series: Materials Science and Engineering* have been peer reviewed through processes administered by the proceedings Editors. Reviews were conducted by expert referees to the professional and scientific standards expected of a proceedings journal published by IOP Publishing.

IOP Conf. Series: Materials Science and Engineering **326** (2018) 012001 doi:10.1088/1757-899X/326/1/012001

A minimal SATA III Host Controller based on FPGA

Hailiang Liu[1,*]

[1]Chengdu Goke Microelectronics Co., Ltd., 5th Floor, Tower B, Oriental Hope Building, 3 Gaopeng Avenue, High-tech Western Zone, Chengdu, Sichuan, China

[*s]hailiangliu@std.uestc.edu.cn

Abstract. SATA (Serial Advanced Technology Attachment) is an advanced serial bus which has a outstanding performance in transmitting high speed real-time data applied in Personal Computers, Financial Industry, astronautics and aeronautics, etc. In this express, a minimal SATA III Host Controller based on Xilinx Kintex 7 serial FPGA is designed and implemented. Compared to the state-of-art, registers utilization are reduced 25.3% and LUTs utilization are reduced 65.9%. According to the experimental results, the controller works precisely and steady with the reading bandwidth of up to 536 MB per second and the writing bandwidth of up to 512 MB per second, both of which are close to the maximum bandwidth of the SSD(Solid State Disk) device. The host controller is very suitable for high speed data transmission and mass data storage.

1. Introduction

In recent years, the area of memory is developing rapidly. As for its low power consumption, good stability and high speed in storage, SSD(Solid State Disk) is gradually occupied the storage market. At present, many SSD products of the main companies in storage area use SATA (Serial Advanced Technology Attachment) interface to transmit data, such as Intel, Toshiba, Samsung and Seagate.

SATA, with a total of 7 signals included with 3 GND signals, a pair of TX differential signals and a pair of RX differential signals, transmitting and receiving with a pair of differential signals, effectively solves the problem of parallel signal transmission crosstalk in high speed digital bus design.

SATA was first proposed by Intel in 2000 to replace the parallel ATA interface at that time. In 2001, SATA 1.0 standard was released and specified rate of data transmission of 150 MB per second. Just one year later, SATA 2.0 standard was born and the transmission speed was up to 300 MB per second. To meet needs of high speed of data transmission and mass data storage, the Serial ATA International Organization released SATA 3.0 standard [1] in 2009 with a speed of 600 MB per second, which was a double rate of the previous generation.

A lot of related research work has been carried out in recent years. Ling yan Fan et al. [2] put forward a novel hardware architecture with embedded AES hardware engine, which made a flash controller simple and reduce the die size. Jianjun Luo et al. [3] implemented a SSD controller with up to five flash channels, which can be configured as RAID0 or RAID5 mode. Abdul Arfan et al. [4] tried to optimize the performance of SATA disks by devising a new cache algorithm that is aware of disk access time. Young-Jin Kim [5] proposed a novel cache algorithm which combined disk access time and access frequency in a more NCQ-friendly way to enhance the I/O performance. From the public information point of view, there are several commercial and academic SATA host controller cores implemented based on FPGA [6,7,8,9,10]. Wu et al. [6,7] implemented SATA HBA core based on

Xilinx Virtex 5 FPGA and SATA revision 1.x in 2008 with a reading bandwidth of up to 111 MB per second and writing bandwidth of up to 102 MB per second. Woods et al.[8] implemented SATA HBA core based on Xilinx Virtex 5 FPGA and SATA revision 2.x in 2012 with a reading bandwidth of up to 279 MB per second and writing bandwidth of up to 242 MB per second. In 2012, Mendon et al. [9] implemented SATA 2.x HBA and one year later, Lehmann et al. [10] implemented SATA HBA based on SATA revision 3.x. The difference of these host controller cores are shown in table 1.

Table 1. Academic SATA cores based on FPGA

	Wu[7]	Woods[8]	Mendon[9]	Weasel[10]	Liu
SATA Rev.	1.x	2.x	2.x	3.x	3.x
NCQ	—	yes	yes	—	—
DMA	yes	FP-DMA	yes	yes	yes
Interface	PLB	FIFO	PLB	FIFO	FIFO
Virtex-5	X	X	X		
Virtex-6			X		
Kintex-7					X
Straix II				X	
Straix IV				X	

This paper designs and implements a minimal SATA III Host Controller based on Xilinx Kintex 7 serial FPGA platform, which is shown in section 2 and section 3. The rest of the paper is organized as follows: Section 2 presents the design of the SATA III Host Controller. The process of implementation is described in Section 3. Section 4 analyzes the experimental results and section 5 concludes this paper.

2. Design phase

In design process, to achieve a minimal SATA III host controller, asynchronous transmitting FIFO and receiving FIFO interface is proposed. Compared to PLB (Processor Local Bus) interface adopted in paper [7, 9], asynchronous FIFO interface is simple to design and flexible to use. What's more, the FIFO interface also cuts down the logic expenditures during implementation phase. Compared to paper [10], which places FIFO before 8b/10b encode and decode, at this time the transmitting data are packaged into FIS (Frame Information Structure) and added with SOF (Start Of Frame), CRC(Cyclic Redundancy Check), EOF(End Of Frame), so our transmitting FIFO could have a relatively shallow depth(the receiving FIFO has a similar explanation), which also saves RAM resource.

In our design, the data width is 32 bits, and the reference clock is 150 MHz. So after 8b/10b encoded and parallel-to-serial, the speed of SerDes is up to 6 Gb per second. Then we choose Xilinx XC7K-325T FPGA as our hardware implementation platform.

In physical layer, the initial of host controller is the key to implement the design. Here figure 1 descr-ibes the initial flow of host controller. When power on or hot plug, host will have a globe reset, then host will send COMRESET, one of three (COMRESET, COMINIT, COMWAKE) OOB (Out Of Band)signals. In the following, host will wait for receiving the COMINIT signal from device. When receiving COMINIT signal, host will send COMWAKE signal to device and wait for receiving COMWAKE signal from device. After receiving COMWAKE for 64 clock period, host will reset Rocket I/O. The reason we do this is to avoid k code error if Rocket I/O doesn't have a reset after host receiving COMWAKE signal. Then host will send D10.2 signal and ALIGNp primitives at 6 Gb per second to device. In order to reduce logical expense, our minimal SATA host controller only supports SATA 3.0 protocol. Device will also send ALIGNp primitives at the same speed to host and the initial of physical layer is completed. Then the link up signal will go high to indicate that the physical link is established, which will be transmitted layer by layer until application layer.

In application layer, in order to verify the performance of Host Controller, one hand we design a random number generator which will write data to SSD device randomly. Correspondingly, an error counter is designed to count the error data reading from SSD device which is not equal to the writing data.

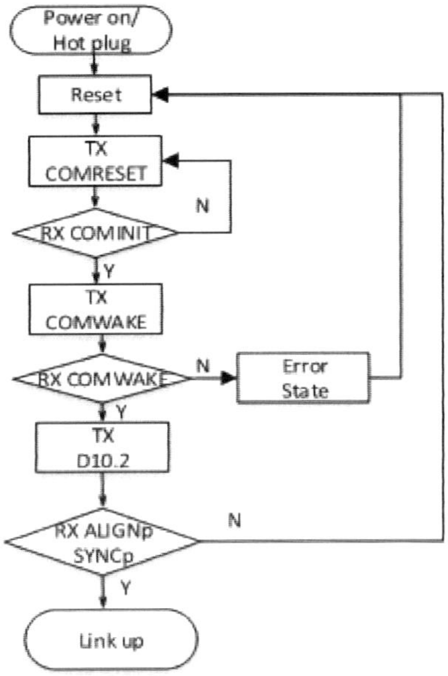

Figure 1. The initial flow of host controller

3. Implementation

Here figure 2 gives the simplest structure of SATA III Host Controller. We can see that the structure could mainly divided into two parts: the writing path and the reading path and each path has four lay-er: application layer, transport layer, link layer and physical layer. As a minimal SATA III Host Contr-oller structure, only the basic function of each layer are implemented.

Application layer provides a FIFO interface to transport layer and responses to answer all interface commands, analyze the device operation commands and do corresponding operation according the protocol. Besides, we add the corresponding logic to monitor the status of writing and reading proce-ss. Based on the Phy initialization status, the host adapter knows whether a device is attached to each of the two ports used in a Master/Slave emulation configuration. Device Control register writes, that have the SRST bit set to one, shall result in the associated Shadow Device Control register being writt-en for each port to which a device is attached. The frame transmission protocol for each associated port executed, results in a Register-Host to Device FIS being transmitted to each attached device. Transport layer presides over serial digital transporting control, mainly including packing control inf-ormation and data into FIS type and transporting to link layer. The another work of transport layer is to manage the conflict between reading and writing, unpack FIS when reading data from device. A FIS is a group of Dwords that convey information between host and device. Primitives are used to define the boundaries of the FIS and may be inserted to control the rate of the information flow.

In link layer, the FISs will comprise a frame with SOF, CRC, EOF. In the process of the transmitting, there will be added ALIGNp, SYNCp, HOLDp, HOLDAp, R OKp, R IPp and so on primitives to con-trol the flow the transmitting. Then the data flow will be scrambled and transmit to physical layer. The Link layer transmits and receives frames, transmits primitives based on control signals from the Trans-port layer, and receives primitives from the Physical layer which are converted to control signals to the Transport layer. The coding scheme used by Serial ATA translates unencoded data and control bytes to characters. The encoded characters are then transmitted by the Physical layer over the serial line where they are received from the Physical layer and decoded into the corresponding byte and control value.

In physical layer, data flow will be encoded in 8b/10b algorithm and transmit to storage device thro-ugh Rocket I/O transceiver with 6 Gb/s. Physical layer provides specified OOB signal detection and transmission, uses OOB signal protocol for initializing the Serial ATA interface, then uses this OOB sequence to execute a pre-defined speed negotiation function. Perform proper power-on sequencing and speed negotiation. So far, all jobs of the writing path are done and the reading path is designed of converse job compared to the writing path.

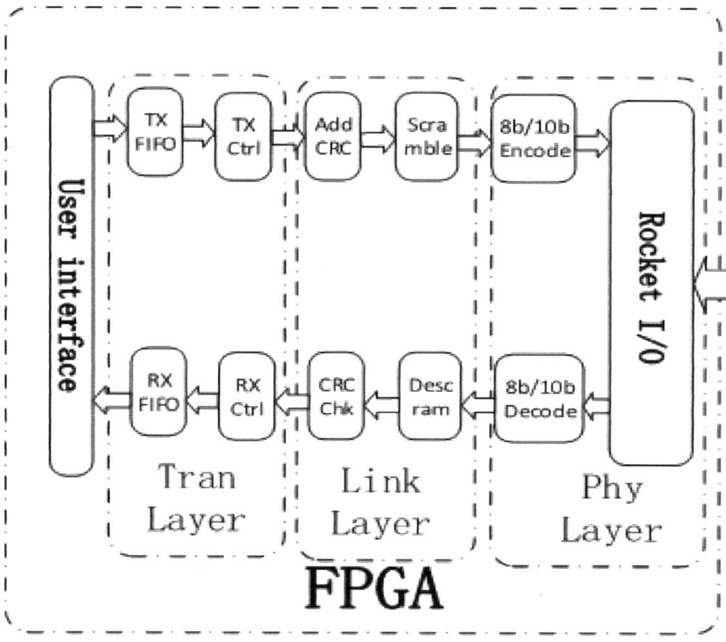

Figure 2. The structure of minimal SATA III Host Controller

4. Results and analysis

As we described before, we implement our SATA III host controller based on Xilinx Kintex XC7K325T FPGA. We choose Samsung Pro 850(MZ-7KE512B) as our storage device, whose reading bandwidth is up to 540 MB per second and writing bandwidth is up to 520 MB per second under ideal condition according to the vendor . As papers[6,7,8,9] are design implementation of SATA I or SATA II HBAs(Host Bus Adapters), the reading bandwidth and writing bandwidth of SATA I, SATA II and SATA III are lack of comparability. Figure 3 is the describe of reading bandwidth and writing band-width of different HBAs[6].

From Figure 3,we can see that the steady reading bandwidth of S2GX:WD Caviar Blue 500G is about 120 MB per second[10], the steady writing bandwidth of S2G:OCZ Vertex3 120G and K7: Samsung Pro 500G is about 535 MB per second, which is close to the maximum value of the storage device. Meanwhile, the reading bandwidth fluctuations of S2G:OCZ Vertex3 120G is a little bigger than K7: Samsung Pro 500G device.

Figure 4 indicates that our writing bandwidth is up to about 510 MB per second, which is close to the maximum value 520 MB per second of the device. How about the FPGA resource utilization? Com-pared to paper[10], we get table 2. It is easy to find that, compared to Weasel on S2GX, our LUTs utilization are reduced 35% and RAMs utilization are reduced from 71 to 2. Compared to Weasel on S4GX, our registers utilization are reduced 25.3% and LUTs utilization are reduced 65.9%.

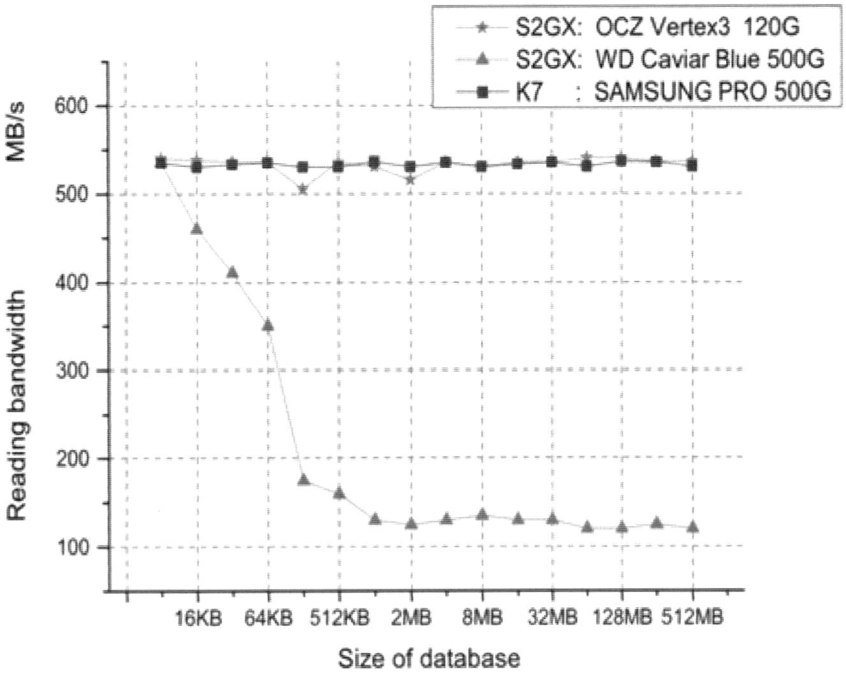

Figure 3. The reading bandwidth of the host controller

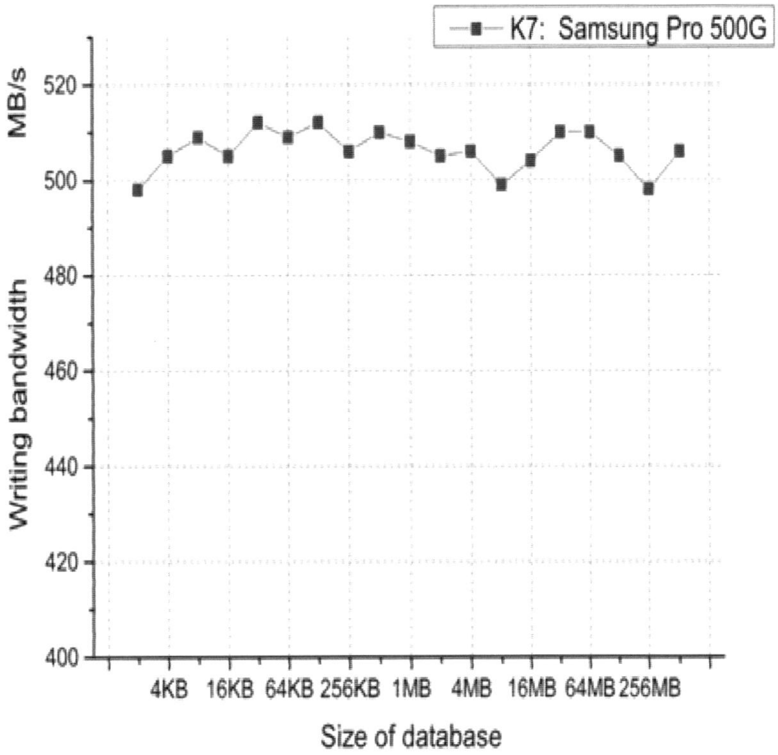

Figure 4. The writing bandwidth of the host controller

IOP Conf. Series: Materials Science and Engineering **326** (2018) 012001 doi:10.1088/1757-899X/326/1/012001

Table 2. FPGA resource utilization compare among different papers

	Reg.	LUTs	RAM	SATA Rev
Weasel on S2GX[10]	1884	2630	71	3.x
Weasel on S4GX[10]	2027	3219	2	3.x
Liu on K7	1617	1940	2	3.x

5. Conclusion

In this express, we design and implement a minimal SATA III host controller based on Xilinx XC7K-325T FPGA and the storage device we choose is Samsung MZ-7KE512B, which is a new kind of high performance SSD. According to the results, the controller works precisely and steady with the reading bandwidth of up to 536M/s and the writing bandwidth of up to 512 MB per second, both of which are close to the max bandwidth of the SSD. Meanwhile, compared to Weasel on S2GX[10], our FPGA resource LUTs utilization are reduced 35% and RAMs utilization are reduced from 71 to 2. Compared to Weasel on S4GX[10], our registers utilization are reduced 25.3% and LUTs utilization are reduced 65.9%. The controller is very suitable for high speed data transmission and huge data storage in SOCs, embedded system or as a SATA III host controller core integrated in FPGA.

Reference

[1] Dell C,Hitachi G, Intel C, Maxim I, Seagate T and Weatern D 2009 *Serial ATA International Organization* vol 3 chapter 7-12 pp 183–535
[2] Lingyan F, Jianjun L, Hailuan Liu and Xuan G 2014 *J. IEICE Electron Express* **11** 20140535
[3] Jianjun L, Lingyan F, Chris T and Xuan G 2014 *J. IEICE Electron Express* **11** 20140419
[4] Abdul A, Young-Jin K and Jin B 2014 *J. IEICE Electron Express* **9** 1707-13
[5] Young-Jin K 2014 J. IEICE Electron Express 11 20140363
[6] Wei W, Hai-Bing S and Qin-Zhang W 2008 Proc. Int. Conf. on Computer Science and Software Engineering (Wuhan) vol 6 (IEEE Press) p 18
[7] Wei W, Hai-Bing S and Qin-Zhang W 2009 Proc. Int. Conf. on Computational Intelligence and Design (Changsha) vol 2 (IEEE Press) p 124
[8] Louis W and Ken E 2012 Proc. Int. Sym. on Field-Programmable Custom Computing Machines (Toronto) vol 20 (IEEE Press) p 220
[9] Ashwin A. M, Bin H and Ron S 2012 Proc. Int. Conf. on Field-Programmable Logic and Applications (Oslo) vol 22 (IEEE Press) p 421
[10] Patrick L, Thomas F, Oliver K, Steffen K, Thomas B. P and Rainer G. S 2013 Proc. Int. Sym. on Field-Programmable Custom Computing Machines (Porto) vol 23 (IEEE Press) p 1

IOP Conf. Series: Materials Science and Engineering **326** (2018) 012002 doi:10.1088/1757-899X/326/1/012002

Effect of Water Cut on Pressure Drop of Oil (D130) -Water Flow in 4"Horizontal Pipe

Mehaboob Basha[1,*], S. M. Shaahid[1] and Luai M. Al-Hems[1]

[1]Centre for Engineering Research- Research Institute, King Fahd University of Petroleum and Minerals, Dhahran, Saudi Arabia

*nbbasha@kfupm.edu.sa

Abstract. The oil-water flow in pipes is a challenging subject that is rich in physics and practical applications. It is often encountered in many oil and chemical industries. The pressure gradient of two phase flow is still subject of immense research. The present study reports pressure measurements of oil (D130)-water flow in a horizontal 4″ diameter stainless steel pipe at different flow conditions. Experiments were carried out for different water cuts (WC); 0-100%. Inlet oil-water flow rates were varied from 4000 to 8000 barrels-per-day in steps of 2000. It has been found that the frictional pressure drop decreases for WC = 0 - 40 %. With further increase in WC, friction pressure drop increases, this could be due to phase inversion.

1. Introduction

Heavy oils represent about one third of the world's hydrocarbon resources, but, their production is associated with huge costs of transportation. Multiphase flow is commonly seen in industrial processes such as pipeline transportation, fluidized beds and power plants. This is the motivation for research in this field. Water desalination using multistage flashing (MSF) process is a simple two-phase flow problem. Blood oxygenation used during open-heart surgery is a multiphase flow process. A typical multiphase oil–water two-phase flow is often encountered in petroleum industries. The measurement of phase flow rates is of particular importance for managing oil production and water disposal and/or water reinjection. There is a need to measure the flow rates of multiphase flow fluids and this is a very challenging process and is of great importance in the petroleum industry.

Pressure (frictional pressure drop) is the key parameter for assessing individual phase (oil and water) flow rates in pipelines. Therefore, it is important to study behaviour of pressure response to characterize two-phase flow in upstream production pipelines. Several articles are available in literature on the two-phase flow of oil and water in pipes.

Xu [1] presented a brief review of oil-water two phase flows in horizontal pipes and highlighted future research trends of oil water pipe flows. Yusuf et al. [2] have presented experimental data of flow patterns, pressure gradient and phase inversion in horizontal oil–water flow in a 25.4 mm acrylic pipe. One of the main findings is the large difference between the pressure gradient results which is attributed to the difference in oil viscosity. The differences between the results become bigger at higher oil velocities. The largest difference in pressure values was observed in flow region where oil is the continuous phase. On the contrary, for dispersed oil in water, the pressure gradient values observed at the same conditions are approximately the same. At low oil velocities, the water velocity required to initiate the transition to non-stratified flow increased as the oil viscosity increased while it decreased at higher oil velocities.

Sotgia et al [3] performed experimental study of oil–water flow in horizontal pipes using mineral oil and tap water of viscosity ratio about 900 and density ratio 0.9. A set of seven different pipes of Pyrex and Plexiglas where used, with diameters ranging between 21 and 40 mm. Pressure drop measurements, flow pattern maps and pictures of the oil–water flow are reported in this article. Talal [4] developed a simple power law pressure gradient correlation for horizontal oil–water separated flow (stratified and dual continuous flows). The new proposed correlation predicts the pressure gradient with higher accuracy. Pressure gradient database was prepared for oil–water flow which includes wide range of operational conditions, fluid properties, pipe diameters and materials.

The formation of water-in-crude oil emulsions during oil production can cause a substantial reduction of the production rates. This occurs due to the high effective viscosity of the emulsion that increases with water content towards the phase inversion point. Knowledge of both the effective viscosity and the phase inversion point is important for the dimensioning of pipelines and equipment as well as for the assessment of production strategies. The effective viscosity of an emulsion can greatly exceed either the crude or the water single phase viscosities. Jose et al [5] made a comparative study of the pipe flow of water-in-crude oil emulsions in a closed loop system (pipe ID2.2cm). The pipe flow of emulsions based on six different crude oils (viscosities from 4.8 to 23.5 mPas) and salt water (3.5% NaCl w/v, pH¼7.3) were investigated experimentally using a small scale flow loop. The effective viscosity of the emulsions as a function of the water fraction was calculated from pressure drop measurements. The point of inversion was observed to be fluid dependent. Pietro et al [6] have studied experimentally the effect of air injection on liquid–liquid core annular flow of very-viscous-oil/water on the pressure drop. They have presented a new data set for pressure drop.

The modeling of the pressure gradient of oil-water flow in pipelines is very crucial. Accurate prediction of pressure gradient will lead to better design of energy efficient transportation systems. Al-Wahaibi and Mjalli [7] developed an artificial neural network (ANN) model with five inputs (oil and water superficial velocities, pipe diameter, pipe roughness, and oil viscosity) to predict the pressure gradient of horizontal oil-water flow based on a databank of around 765 measurements collected from the open literature. Statistical analysis showed that the ANN model has an average error of 0.30%. Hasanvand and Berneti [8] have also used artificial neural networks (based on 600 data set of Persian Gulf oil) in their study to obtain the oil flow rate as output measurement, The input variables included temperatures and line pressures. Tan et.al [9] conducted experiments (in a 50 mm diameter horizontal pipe) to measure the individual phase rates of oil-water two phase flow using a Conductance Ring Coupled Cone (CRCC) meter. The accuracy of this method in measuring oil flow rate is 2.3% and for water flow rate is 4.8%. The proposed method improves the accuracy of the measurement by measuring the water holdup in the annular channel.

An experimental investigation to study the slip (holdup) phenomenon between phases in water-oil two phase flows in horizontal pipes was conducted out by Jing et al. [10]. Emphasis was placed on the effects of input fluids flow rates, pipe diameter (0.05m and 0.025 m) and viscosities of oil phase on the slip. Results showed that at low input flow rate, there is a large deviation on the holdup between two flow systems with different oil viscosities and the deviation becomes gradually smaller with increased input water flow rate.

In the light of the above literature survey, there is no work available on pressure drop measurements of oil (D130)-water two-phase flow in horizontal 4 inch diameter stainless steel pipe at different flow conditions. This is the driving force for the present experimental study and it focuses on the effect of flow rates, water-cuts on pressure drop measurements of oil (D130)-water two-phase flow.

In this study, efforts have been made to present pressure drop measurements of oil (D130)-water two-phase flow in a horizontal 4 inch diameter stainless steel pipe at different flow conditions. Experiments were conducted for different water cuts; 0%, 20%, 40% 60% 100%. Inlet oil-water flow rates were varied from 4000 to 8000 barrels-per-day (BPD).

2. Experimental setup

The Oil-water two phase experiments were carried out at the multi-phase laboratory of King Fahd University of Petroleum and Mineral, Dhahran, Saudi Arabia.

The layout of the flow loop is presented in figure 1. Experimental set-up consists of: four centrifugal variable speed pumps [2 pumps for water (WP) and 2 pumps for oil, (OP)], 4 inch stainless loop, a horizontal separator tank (WOST), which serves as a storage tank, two level indicators for oil and water each. The loop is mounted on swinging platform (angle can be varied from 0° - 30°). The fixing of loop at a given angle is done by flexible connection (FC).

The instruments of the loop include: a turbine type oil flow meter (OFM), a turbine type water flow meters (WFM), line pressure transmitter (LPT), two flow differential pressure transmitters (DPT1 and DPT2). More information of the loop components and instruments is given in table 1.

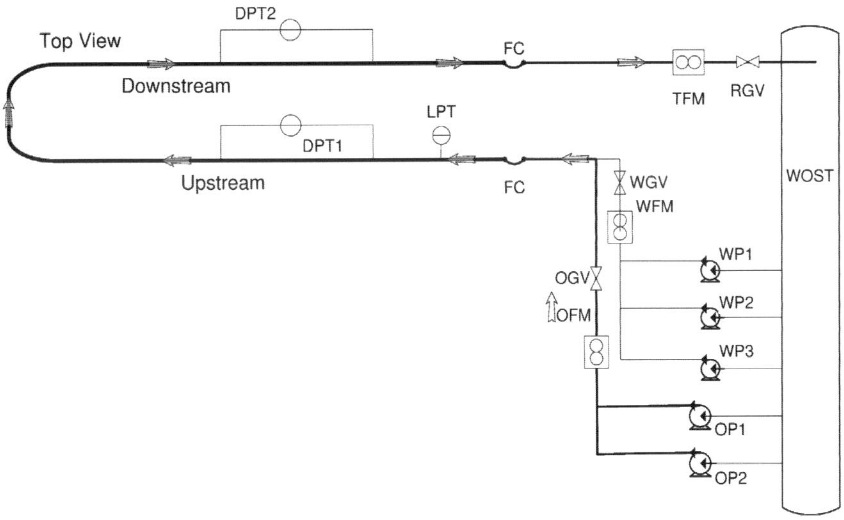

Figure 1. Schematic layout of the oil-water multiphase flow loop

Table 1. Details of equipment of the flow loop.

Items	Manufacturer	Model	Capacity/Range	Accuracy/Error
Four pumps (two water, two oil)	NEWAR FLOW SERVE	50-32CPX200	35 m^3/hr	-
Two turbine flow meter	Omega	EF10	±10 m/s	±1.0 %
Line pressure gauge	ROSEMOUNT	AOB-20	0-7 bar	±0.25%
DPT1	ROSEMOUNT	300S2EAE5M9	0-70 inches of water	±0.1%
DPT2	ROSEMOUNT	300S2EAE5M9	0-12 inches of water	±0.1%

3. Experimental procedure

Initially, experiments were conducted for water-only and oil-only single phase (in 4inch pipe) to validate the pressure drop measurements against available empirical models, and to ascertain effectiveness of pressure transmitters and flow meters of the loop.

In this regard, water was pumped in the loop using centrifugal pumps. The desired volume flow rate was obtained by varying speed of pumps through variable speed drives and also by regulating oil globe valve (OGV) and water globe valve (WGV) of oil and water flow streams respectively. The flow rates on the discharge line of the pumps were measured by Turbine flow meters. The required outlet pressure (e.g. 1 bar or 2 bars) of the loop is set by throttling the Return gate valve (RGV, figure1).

The experiments were carried out for a given flow rate and pressure drop measurements were made at different locations of the loop as shown figure 1. After achieving the steady state flow condition, differential pressure drops are recorded across 3m (DPT1). The experimental data was recorded using CR 1000 data logger. The above procedure was repeated for oil-only flow experiments.

The friction factor was calculated using the pressure drop data by using Eq. (1) and compared with Eq. (2) and Eq. (3).

$$f = \frac{\Delta P}{L}\frac{2D}{\rho v^2} \tag{1}$$

ΔP	Pressure drop (Pa).	v	Average velocity of the fluid (m/s)
L	Distance between the two pressure taps (m)	ε	Pipe roughness (m)
D	Hydraulic diameter of the pipe (m)	Re	Reynolds number
ρ	Fluid density (Kg/m^3)		

$$f = 0.3164\,Re^{-1/4} \tag{2}$$

$$\frac{1}{\sqrt{f}} = -2\log\left[\frac{\varepsilon/D}{3.7} - \frac{5.02}{Re}\log\left[\left(\frac{\varepsilon/D}{3.7}\right) + \frac{13}{Re}\right]\right] \tag{3}$$

The Zigrang & Sylvester 1985 correlation defined in equation (3) above can also be used to determine the turbulent friction factor.

The comparison of experimentally obtained friction factor (Eq. 1) with the friction factors calculated by using Blasius correlation and Zigrang & Sylvester correlations is shown in the Figure 2. A good agreement has been noticed specifically with the Blasius friction factor (Eq. 2).

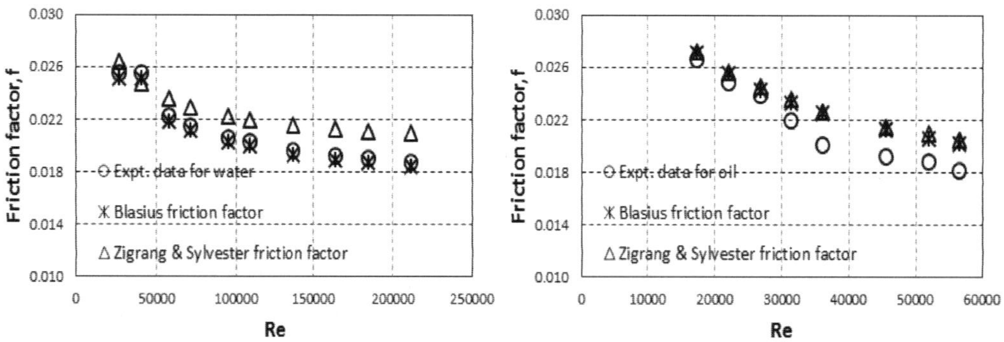

Figure 2. Friction factor comparisons with Blasius correlation and Zigrang & Sylvester correlations for oil-water flow

The variation single phase water and oil friction factor against Reynolds number is shown in figure 2. The friction factor has observed to decrease with increase in velocity. A good agreement is can be noticed between the experimental data and established theoretical relation.

For a given oil-water multiphase flow, the required flow rate and water cut is obtained by varying the speeds of the oil and water pumps. After the required water cut and flow rates are achieved, pressure drop [across 3m (DPT1)] measurements were made. The above procedure was repeated for different water cut ratio 0 to 100%. Inlet oil-water flow rates were varied from 4000 to 8000 barrels-per-day (BPD).

4. Results and discussions

Oil-water multiphase flow experiments were conducted for different water cut ratios (0%, 20%, 40%, 60% and 100%). Inlet oil-water flow rates were varied from 4000 to 8000 barrels-per-day (BPD).

4.1. Effect of water-cut on oil-water pressure drop for different flow rates

Figure 3a shows the effect of water cut for different flow rates on pressure drop. It can be seen from figure 3a that at a given flow rate the pressure drop decreases from WC = 0 to WC 40 % whereas for WC 40 % to WC 60 %, friction pressure drop has been found to increase. This could be due phase inversion. However, for WC = 100%, frictional pressure drop is lower as compared to frictional pressure at WC = 0%. This is due to lower viscosity of water. Also, it can be observed from figure 3a, that at any given WC, the frictional pressure drop increases with increase in flow rate. For a given water cut WC = 40, increasing in BPD from 4000 to 6000, percentage increase in frictional pressure drop is about 150%.

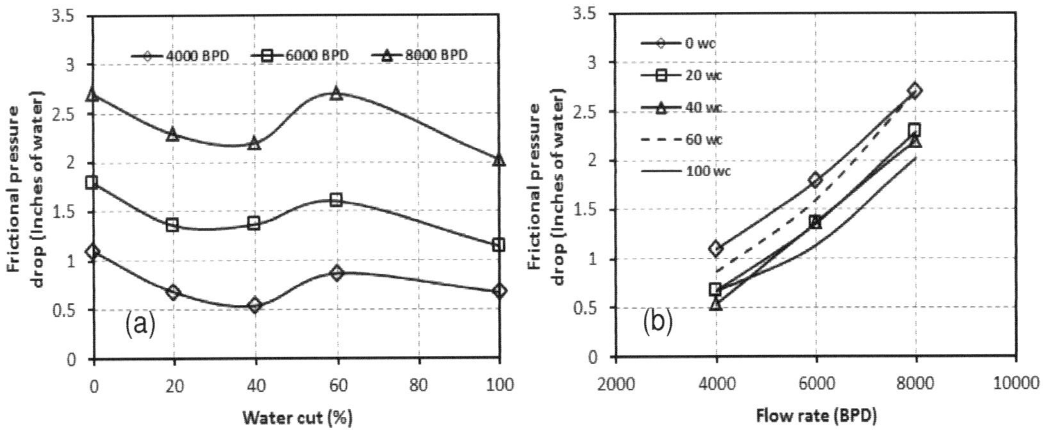

Figure 3. Frictional pressure drop behavior. (a) Effect of water cut on pressure drop for different flow rates, (b) Effect of flow rate on pressure drop for different water cuts (0° case)

4.2. Effect of flow rate on oil-water pressure drop for different water-cuts

Figure 3b shows the effect of flow rate on frictional pressure drop for different water cuts. As stated earlier, it can be seen from figure 3b, pressure drop increases with flow rate and WC. The frictional pressure has been found to increase linearly with respect flow rate. However, effect of water cut on frictional pressure drop is not linear. For a given flow rate 6000 BPD, increasing in water cut from WC 20 to 40, percentage increase in frictional pressure drop is about 17%.

5. Conclusions

In the present study, pressure drop measurements of oil (D130)-water two-phase flow in a horizontal 4″ diameter stainless steel pipe at different flow conditions were made. Experiments were conducted for different water cut ratios (0%, 20%, 40%, 60% and 100%). Inlet oil-water flow rates were varied from 4000 to 8000 barrels-per-day (BPD). Measured pressure drops and friction factor of single phase oil and single phase water were compared with existing empirical relations and good agreement was found.

For a given flow rate, the frictional pressure drop has been found to decrease for WC = 0 to WC 40 %. Further increase in WC, friction pressure drop has been found to increase. The frictional pressure has been found to increase linearly with respect flow rate. However, effect of water cut on frictional pressure drop is not linear.

6. Nomenclature

A	Cross-sectional area of pipe $[m^2]$	ρ_w	Density of water [kg/m^3]
BPD	Barrel per day	ρ_o	Density of oil [kg/m^3]
Dh	Diameter of the pipe [m]	μ_o	Viscosity of oil [$Pa.s$]
f	Friction factor	μ_w	Viscosity of water [$Pa.s$]
ID	Inner diameter [m]	ΔP	Pressure drop [$inch\ H_2O, pa$]
L	Length of the pipe [m]	$\frac{\Delta P}{\Delta L}$	Pressure gradient [$\frac{Pa}{m}, \frac{inch\ H_2O}{m}$]
Re	Reynolds's number		

Acknowledgments
The authors acknowledge the support of Center for Engineering Research, Research Institute, King Fahd University of Petroleum and Minerals for this research work.

References

[1] Xu X. X. (2007), Study on oil water two phase flow in horizontal pipe lines, *Int. J. Petr. Sci.* Engg., 59, pp. 43–58.

[2] Yusuf N., Y. Al-Wahaibi., T. Al-Wahaibi., A. Al-Ajmi and A.S. Olawale (2012). Effect of oil viscosity on the flow structure and pressure gradient in horizontal oil–water flow, *Chemical Engineering Research and Design*, v. 90, pp. 1019-1030.

[3] Sotgia G., P. Tartarini, and E. Stalio (2008). Experimental analysis of flow regimes and pressure drop reduction in oil–water mixtures. *International Journal of Multiphase Flow,* v. 34, pp. 1161–1174.

[4] Talal Al-Wahaibi (2012). Pressure gradient correlation for oil–water separated flow in horizontal pipes. Experimental Thermal and Fluid Science, v. 42, pp. 196–203.

[5] Jose P., B. Pettersen, and O.Nydal (2013). Pipe flow of water-in-crude oil emulsions: Effective viscosity, inversion point and droplet size distribution. *Journal of Petroleum Science and Engineering*, v. 10, pp. 35–43.

[6] Pietro P., D. Strazzaa, and G. Sotgia (2009). Very-viscous-oil/water/air flow through horizontal pipes:Pressure drop measurement and prediction. *Chemical Engineering Science*, v. 64, pp. 1136 – 1142.

[7] T. Al-Wahaibi and F. S. Mjalli (2014), Prediction of Horizontal Oil-Water Flow Pressure Gradient Using Artificial Intelligence Techniques, *Chem. Eng. Commun.*, vol. 201, no. 2, pp. 209–224.

[8] M. Hasanvand and S. M. Berneti (2015), Predicting Oil Flow Rate due to Multiphase Flow Meter by Using an Artificial Neural Network, *Energy Sources, Part A Recover. Util. Environ. Eff.*, vol. 37, no. 8, pp. 840–845.

[9] C. Tan, W. Dai, H. Yeung, and F. Dong (2015), A Kalman estimation based oil-water two-phase flow measurement with CRCC, *Int. J. Multiph. Flow*, vol. 72, pp. 306–317.

[10] Jing-yu X., Ying-sing W., Fei-fei F., Ying C., and Dong-hui-L (2008). Experimental investigation on the slip between oil and water in horizontal pipes, *Journal of Thermal and Fluid Science*, 33, pp. 178-183.

IOP Conf. Series: Materials Science and Engineering **326** (2018) 012003 doi:10.1088/1757-899X/326/1/012003

Stochastic Investigation of Natural Frequency for Functionally Graded Plates

P K Karsh[1], T Mukhopadhyay[2] and S Dey[1,*]

[1]Mechanical Engineering Department, National Institute of Technology Silchar, India
[2]Department of Engineering Science, University of Oxford, United Kingdom

*Corresponding author's e-mail: infodrsudip@gmail.com

Abstract. This paper presents the stochastic natural frequency analysis of functionally graded plates by applying artificial neural network (ANN) approach. Latin hypercube sampling is utilised to train the ANN model. The proposed algorithm for stochastic natural frequency analysis of FGM plates is validated and verified with original finite element method and Monte Carlo simulation (MCS). The combined stochastic variation of input parameters such as, elastic modulus, shear modulus, Poisson ratio, and mass density are considered. Power law is applied to distribute the material properties across the thickness. The present ANN model reduces the sample size and computationally found efficient as compared to conventional Monte Carlo simulation.

1. Introduction

Functionally graded materials (FGM) belong to the category of advanced materials, which contain two different materials i.e. metal and ceramic. The mechanical, chemical and thermal properties of FGM are totally different from the individual parent materials [1]. Traditional methods cannot be used to dissolve two or more alloying elements due to thermal equilibrium limit. FGM is an advances of composite, in which material properties are varying continuously throughout the depth [2]. Figure 1 shows the geometric view of FGM cantilever plate, in which top of the plate is ceramic rich, whereas bottom is metal rich. Composites have some limitations i.e. it fails under extreme working condition due to delamination of layer, but in FGM there is no layer, so it can be used for long working condition [3]. Many researchers worked on free vibration analysis by using different theories, and models. The ANN model is developed for analysis of stochastic natural frequency analysis of laminated composite plates [4]. A combined application of FEM and ANN are utilised to found the effect of laminating stacking sequence of composite on ballistic limit [5]. First order deformation theory is applied to static and free vibration analysis of FG carbon nanotube skew plate with considering different skew angle, aspect ratio, thickness to width ratio, and boundary conditions [6]. Mori–Tanaka scheme, Power law, and third-order deformation theory are used to investigate the effect of loading condition, volume-fraction exponent, skew angle, and aspect ratio on deflection, natural frequency, and critical buckling load of FG skew plate [7]. Finite element method is applied to determine the natural frequency and mode shape of FGM at different boundary conditions, and different value of power law index [8]. 3D-shell model and discrete double directors shell model (DDDSM) are developed to analysis of free vibration of FGM plates [9]. State-space based differential quadrature method (SSDQM) is used to analysis of natural frequency of FGM plate. ANN is also applied for comparative study for different boundary conditions, thickness of the annular plate, and material property graded index [10]. In past, various studies on stochastic analysis of composite have been conducted [11-16]. The present study is computationally investigated on stochastic natural frequencies of the FGM plates by using ANN approach.

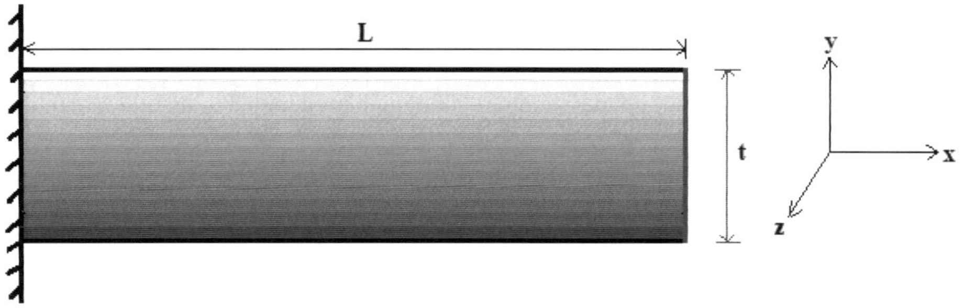

Figure 1. Geometry of FGM cantilever plate

2. Mathematical Formulation

In Cartesian coordinate system (x, y, z) as shown in Figure 1, assume (x, y) is the mid plane of the reference plane. The displacement can be expressed as

$$u(x,y,z) = u_0(x,y) - z\theta_x(x,y)u(x,y) \quad (1)$$

$$v(x,y,z) = z_0(x,y) - z\theta_y(x,y)v(x,y) \quad (2)$$

$$w(x,y,z) = w_0(x,y) = w(x,y) \quad (3)$$

Where *u, v*, and *w* are the displacement components in the x, y and z direction respectively. u_0, v_0 and w_0 are the displacement at the mid plane and θ_x , θ_y are the rotations of cross sections along the x and y axes. Using Hamilton's principles, the dynamic equilibrium equation employing Lagrangian can be represented as

$$L_f = T - U - W \quad (4)$$

Where *T, U* and *W* denotes total kinetic energy, total strain energy and total potential of the applied load, respectively. The stochastic dynamic equilibrium equation considering free vibration can be expressed as

$$[M(\hat{\omega})]\{\ddot{\delta}\} + [K(\hat{\omega})]\{\delta\} = 0 \quad (5)$$

Where $(\hat{\omega})$ denotes the degree of stochasticity and {δ} is global displacement vector and {F} is Global vector of externally applied load, $[M(\hat{\omega})]$= Global mass matrix, $[K(\hat{\omega})]$= Global stiffness matrix.

If '*M*' denotes a function of the material properties can be expressed as,

$$\boldsymbol{M} = \boldsymbol{M}_0 + \boldsymbol{M}_{-1}\boldsymbol{T}^{-1} + 1 + \boldsymbol{M}_1\boldsymbol{T} + \boldsymbol{M}_2\boldsymbol{T}^2 + \boldsymbol{M}_3\boldsymbol{T}^3 \quad (6)$$

Here, M_0, M_{-1}, M_1, M_2, M_3 are the coefficients of temperature *T* in Kelvin. The effective material properties (*M*) are obtained using the power law distribution,

$$M(t) = M_m + (M_c - M_m)\left[\frac{z}{t} + \frac{1}{2}\right]^p \quad (7)$$

Where '*c*' and '*m*' designates the equivalent values at the external surface (ceramic rich) and internal surface (metal rich) of functionally graded plates. Here t and p denote the thickness and power law exponent or material property graded index, respectively and, $z = (t/2)$.

3. Stochastic Approach using ANN Model

In the present study, the most sensitive input parameters, namely elastic modulus, shear modulus, Poisson ratio and mass density are considered for combined variation cases. Hence, in this paper, the material properties of functionally graded plate are considered as input parameters for stochastic analysis of natural frequencies. A combined variations in input parameters(s) can be considered as follows:

Combined variation of material properties

$$g\{ E_1(\varpi), G_{12}(\varpi), \mu(\varpi), \rho(\varpi) \} = \{\Phi1 (E_1(1), E_1(2), \ldots E_1(i)), \Phi2 (G_{12}(1), G_{12}(2), \ldots G_{12}(i)), \Phi3 (\mu(1), \mu(2) \ldots \mu(i)), \Phi4 (\rho(1), \rho(2), \ldots \rho(i))\} \quad (8)$$

Where E_1, G_{12}, μ, and ρ are longitudinal elastic modulus, longitudinal shear modulus, Poisson's ratio, and mass density respectively. Weights, bias, and an activation function are the main unit of the ANN. Figure 2 shows the basic elements of ANN, where input is given to each neuron, and weight function is multiplied with each input. The bias (B_i) is a nonzero value, which is added with the summation of inputs and corresponding weight, given as

$$R_i = \sum_{i=1}^{n} w_{ij} y_i + B_i \quad (9)$$

The summation R_i is transformed using activation function also called transfer function, Z_i represents a value called the unit's "activation"

$$Z_i = f(R_i) \quad (10)$$

4. Results and Discussion

In this paper, probability density function (PDF) plot is considered as the benchmark of the bottom line results. First three natural frequencies are considered to validate the results proposed by ANN model with respect to the direct Monto Carlo simulation. A comparative study between the results obtained from ANN model and MCS is carried out. Probability density function (PDF) for first three natural frequencies (rad/s) due to combined variation of variables, elastic modulus, shear modulus, Poisson's ratio, and mass density for ANN model and MCS model is shown in Figure 2. ANN model shows better results with higher number of sample size as shown in the figure 2, in which sample size N=1024 gives results almost same to MCS. Scatter plot for first three natural frequencies (rad/s) for ANN model with respect to original finite element model considering sample size (N) = 1,024 for combined variation of elastic modulus, shear modulus, Poisson's ratio, and mass density is shown in Figure 3. It is evident from Figure 3, the results obtained from the ANN model are close to the MCS, which gives almost same range of natural frequency for fundamental natural frequency.

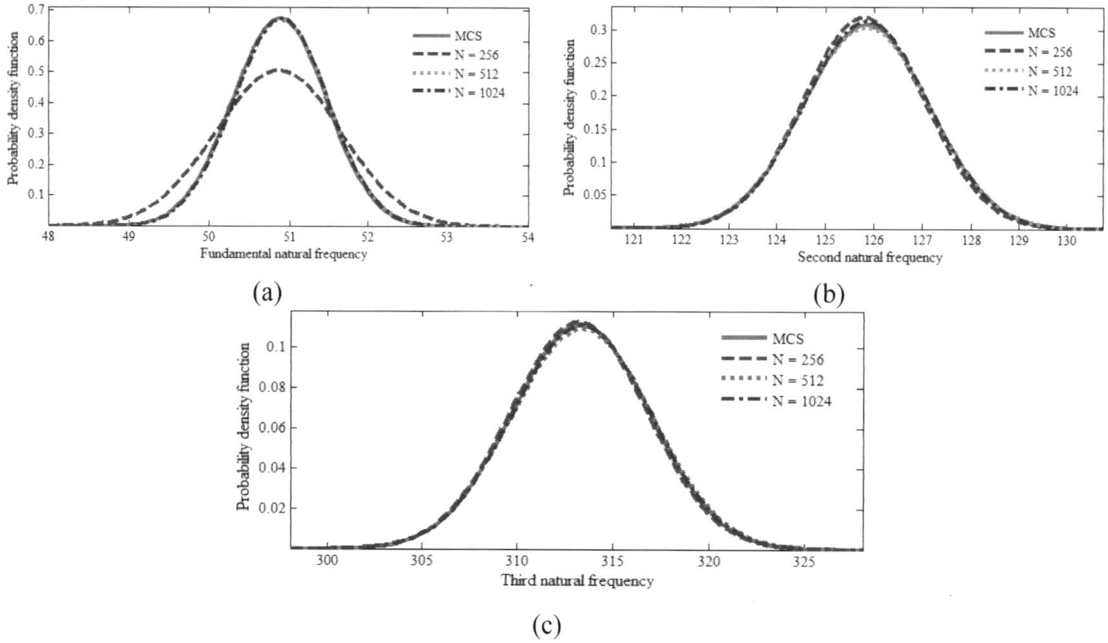

Figure 2. Probability density function for first three natural frequencies (rad/s) due to combined variation of variables for ANN with different sample size and MCS.

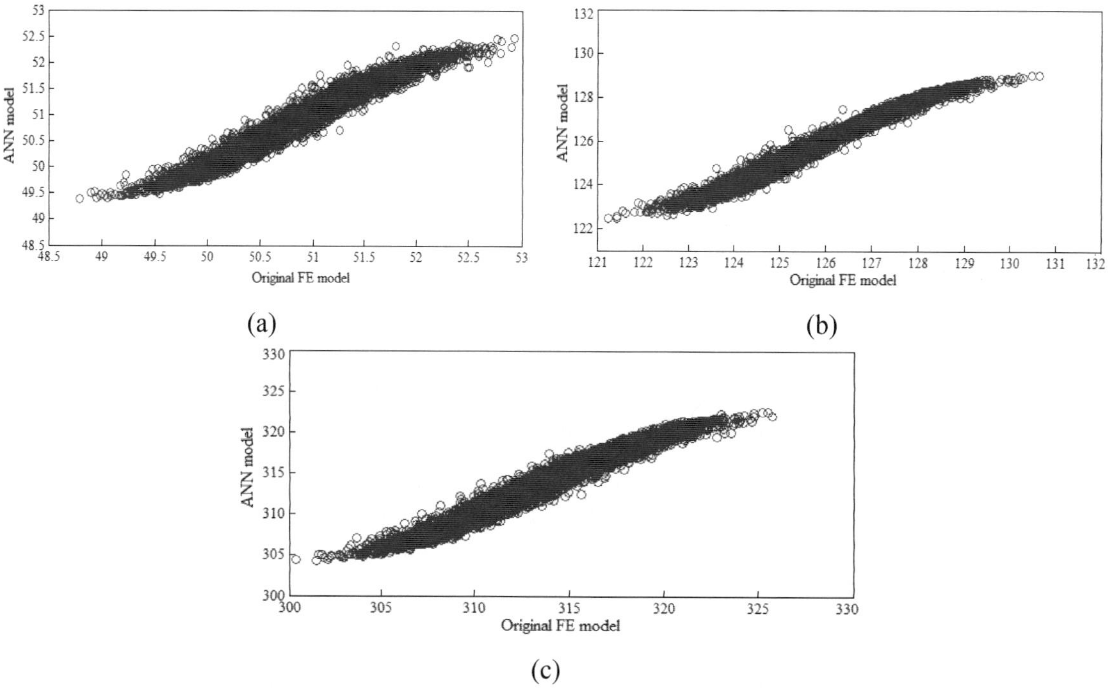

Figure 3. Scatter plot of first three natural frequencies (rad/s) for ANN model with respect to original finite element model considering N_{samp} = 1024 for combined variation of elastic modulus, shear modulus, Poisson's ratio, and mass density.

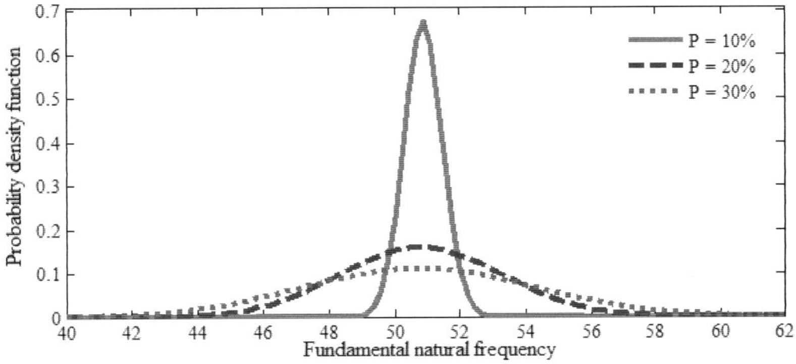

(a)

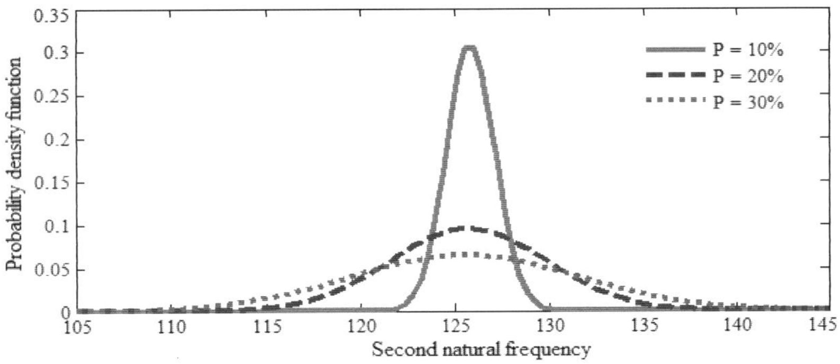

(b)

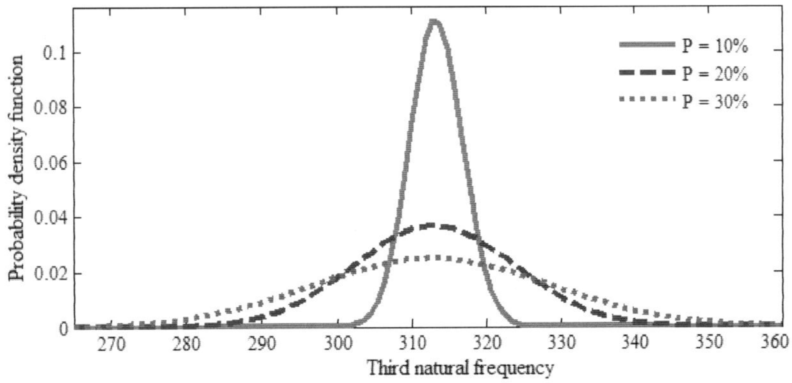

(c)

Figure 4. Probability density function for first three natural frequencies (rad/s) due to combined variation of combined variables, elastic modulus, shear modulus, Poisson's ratio, and mass density at different percentage of variation (P).

Figure 4 shows the probability density function for first three natural frequency due to combined variation at different percentage of variation (P) i.e. 10%, 20%, and 30%. It is clear from the Figure 4, with increase in the percentage of variation (P), stochasticity of all three natural frequencies is also increases. For all natural frequencies, stochasticity in output quantity of interest (QoI) is minimum, when percentage of variation is 10% as compared to 20% and 30%.

5. Conclusions

The novelty of the present study includes the incorporation of artificial neural network in conjunction to stochastic finite element algorithm for FGM plates. Both individual and combined variations of input variables are demonstrated. From the analyses presented in this article it is found that as the percentage of variation of input parameters increases, the sparsity of the stochastic output natural frequencies also increases while no notable variation of stochastic mean value of respective natural frequencies is identified. The computational time and cost is reduced by using the present ANN approach compared to traditional Monte Carlo simulation method. This ANN based approach can be extended to deal with more complex systems in future.

Reference

[1] Mirtalaie S.H., Hajabasi M.A. and Hejripour F. "Free Vibration Analysis of Functionally Graded Moderately Thick Annular Sector Plates Using Differential Quadrature Method" 2012 *Applied Mechanics and Materials* **110-116** 2990-98

[2] Dey S, Mukhopadhyay T and Adhikari S. "Stochastic free vibration analyses of composite shallow doubly curved shells – a Kriging model approach" 2015 *Composite B Eng* **70** 99-112

[3] Vimal Jyoti, Srivastava R K, Bhatt A D and Sharma Avadesh K. "Free vibration analysis of moderately thick functionally graded skew plates" 2014 *Engg. Solid Mechanics* **2** 229-38

[4] Dey S, Mukhopadhyay T, Spickenheuer A, Gohs U and Adhikari S "Uncertainty quantification in natural frequency of composite plates - An Artificial neural network based approach"2016 *Advanced Composites Letters* **25** 2 43-48

[5] Artero-Guerrero J.A., Pernas-Sanchez J., Martin-Montal J, Varas D and Lopez-Puente J. "The influence of laminate stacking sequence on ballistic limit using a combined Experimental/FEM/Artificial Neural Networks (ANN) methodology" 2017 *Composite Structures* doi. 10.1016.

[6] Garcia-Macias E., Castro-Triguero R., Flores E. I. S, Friswell M. I. and Gallego R. "Static and free vibration analysis of functionally graded carbon nanotube reinforced skew plates" 2016 *Composite Structures* **140** 473–90

[7] Taj G. and Chakrabarti A. "Static and Dynamic Analysis of Functionally Graded Skew Plates" 2013 *Journal Engineering Mechanics* **139** 848-57

[8] Ramu I and Mohanty S.C "Modal analysis of Functionally Graded material Plates using Finite Element Method" 2014 *Procedia Materials Science* **6** 460-67

[9] Wali M., Hentati T., Jarraya A. and Dammak F. "Free vibration analysis of FGM shell structures with a discrete double directors shell element" 2015 *Comp. Struc* **125** 295-303

[10] Jodaei, M. Jalal and M.H. Yas "Free vibration analysis of functionally graded annular plates by state-space based differential quadrature method and comparative modelling by ANN" 2012 *Composites Part B* **43** 340-53

[11] Dey S, Mukhopadhyay T. and Adhikari S. "Stochastic free vibration analysis of angle-ply composite plates - A RS-HDMR approach" 2015 *Composite Structures* **122** 526–36

[12] Dey S, Mukhopadhyay T, Haddad Khodaparast H. and Adhikari S. "Stochastic natural frequencies of composite conical shells" 2015 *Acta Mechanica* **226** 8 2537-53

[13] Dey S, Mukhopadhyay T., Spickenheuer A., Adhikari S. and Heinrich G. "Bottom up surrogate based approach for stochastic frequency response analysis of laminated composite plates" 2016 *Composite Structures* **140** 712-27

[14] Dey S, Mukhopadhyay T., Sahu S.K. and Adhikari S. "Effect of cutout on stochastic natural frequency of composite curved panels" 2016 *Composites Part B: Engineering* **105** 188-202

[15] Dey S., Mukhopadhyay T., Sahu S.K., Li G., Rabitz H. and Adhikari S. "Thermal uncertainty quantification in frequency responses of laminated composite plates" 2015 *Composites Part B* **80** 186-97

[16] Dey S. and Karmakar A. "Free vibration analyses of multiple delaminated angle-ply composite conical shells – A finite element approach" 2012 *Composite Structures* **94** 2188–96

IOP Conf. Series: Materials Science and Engineering **326** (2018) 012004 doi:10.1088/1757-899X/326/1/012004

Study on Fatigue Performance of Composite Bolted Joints with Bolt-Hole Delamination

M J Liu[1,*], S Yu[1] and Q Y Zhao[1]

[1]State Key Laboratory of Structural Analysis for Industrial Equipment, School of Aeronautics and Astronautics, Dalian University of Technology, Dalian 116024

*liumj@dlut.edu.cn

Abstract. Fatigue performance of composite structure with imperfections is a challenging subject at present. Based on cohesive zone method and multi-continuum theory, delamination evolution response and fatigue life prediction of a 3D composite single-lap joint with a bolt-hole have been investigated through computer codes Abaqus and Fe-safe. Results from the comparison of a perfect composite bolted joint with another defect one indicates that a relatively small delamination damage around the bolt hole brings about significant degradation of local material performance. More notably, fatigue life of stress concentration region of composite bolted joints is highly sensitive to external loads, as an increase of 67% cyclic load amplitude leads to an decrease of 99.5% local fatigue life in this study. However, the numerical strategy for solving composite fatigue problems is meaningful to engineering works.

1. Introduction

Throughout the last four decades the application of composite materials in engineering structures has been steadily growing from sports equipment and high performance racing cars, to helicopters and commercial aeroplanes. These large-scale mechanical structures generally consist of thousands of small parts which are fastened together to create the final assembly. Different from metals, composite material features a pronounced notch sensitivity which especially develops under quasi-static loading. However, composite bolted joints stand out due to their excellent fatigue resistance and a residual strength as a consequence of inherent material softening mechanisms. What's more, bolted connection is efficient, reliable and economical. Thus bolted joints are one of the most common elements in construction and machine design using composite materials. Grinding and drilling holes are the most frequently employed operations of secondary machining for composite materials owing to the need for structure joining. One of the major problems during grinding and drilling is the delamination damage caused by the tool thrust. The delamination causes concentrated stresses and compromises the mechanical properties of the finished part but, more importantly, it may lead to reduced load-bearing capacity and fatigue failure.

The origin of this study lies within the framework of an industrial agreement to study the fatigue performance of composite bolted joints with delamination around its bolt hole. There are three kinds of approaches for the study of composite delamination under fatigue loading: phenomenological fatigue damage models, semiempirical models based on experiments and numerical simulation approach using finite element method (FEM).Phenomenological fatigue damage models predicts fatigue damage growth according to residual stiffness or residual strength whose evolution rules are based on experiment phenomenons.

Phenomenological models are commonly applied in Engineering, but they model the fatigue behavior of composite materials by a single continuum, occasionally anisotropic, whose constitutive constants are chosen in such a way that the behavior of the single continuum provides an approximation to the behavior which takes no consideration of inner damage mechanism of each constituent material and their interactions. There are a vast amount of literatures providing phenomenological fatigue damage models for many kinds of composite materials [1-5]. Semiempirical models study fatigue process using some parameters from experiment according to theory of continuum damage mechanics or microscopic damage mechanics. Semiempirical models based on continuum damage mechanics establish the damage evolution equation on the basis of the thermodynamic potential function and the dissipative potential energy function which is related to the generalized force. And the damage varies in accordance with the generalized force, which could be derived from the rate of change of the dissipative potential energy function [6-7]. Semiempirical models based on microscopic damage mechanics start from local stress state, combine the typical damage units, such as the micro-cavity, the micro-cracks and the shear band, and then give the rules of deforming and damage evolution for the combined typical damage units. Subsequently, we can reflect the macroscopic damage mechanism to meso mechanical behavior through the mechanical averaging methods, which makes it possible to predict directly the fatigue life and other characteristics of composite structures [8-10]. Numerical methods are always effective and convenient to give reasonable solutions for most engineering problems with various initial conditions and boundary conditions, among which FEM is most commonly used. Generally, the process of composite fatigue includes stiffness degeneration, Stress redistribution, damages accumulation, delamination growth and macroscopic failure. FEM relates variation of stiffness with fatigue process and then gives stress analysis, failure analysis and material performance degradation of structures [11-13]. Of all the three kinds of methods, Numerical methods are more cost-effective and time-consuming, thus present study makes an attempt to evaluate fatigue performance of composite bolted joints with bolt-hole delamination utilizing commercial codes.

2. Numerical approach

In this paper, commercial codes are used to analyze fatigue performance of composite bolted joints with bolt-hole delamination. Two parts are considered, the first part is the delamination initiation simulation of composite bolted joints by using commercial software abaqus, and the second part is fatigue analysis of the composite bolted joints with initial delamination damage by using commercial software fe-safe. Simulation of the first part utilized cohesive zone model approach and analysis of the second part is based on multicontinuum theory (MCT). Results of the first part are initial states of the second part.

2.1. Cohesive zone model approach

For the production of composite structures, machining operations like drilling are frequently needed in composite structures, as the use of bolts, rivets or screws is required to join the parts. Generally, machined parts have poor surface appearance due to material anisotropy and low interlaminar shear strength. One of the most common problems is the elamination damages around the drilling holes, which can decrease the performance of the composite structure. For this issue in simulation works, abaqus applied a cohesive zone model (CZM) approach which has given satisfactory results for simulation of the initiation and propagation of delamination damage. Cohesive damage zone models relate tractions, τ , to displacement jumps, $\overline{\Delta}$, at an interface where a crack may occur. Damage initiation is related to the interfacial strength τ^0.When the area under the traction-displacement jump relation is equal to the fracture toughness, G_c , the traction is reduced to zero and new crack surfaces are formed. If a linear softening law is used, the new crack surfaces are completely formed when the displacement jump is equal to, or greater than, the final displacement jumps $\overline{\Delta}^f$ (see Fig.1):

$$\begin{cases} \tau=\left(1-\bar{d}\right)\tau^{0} & \bar{\Delta}<\bar{\Delta}^{f} \\ \tau=0 & \bar{\Delta}\geq\bar{\Delta}^{f} \end{cases} \tag{1}$$

In the cohesive damage model, the damage variable $\bar{d}$ describes the density of microcracks of are presentative element surface, when $\bar{d}=0$, the element has no damage, when $\bar{d}=1$, the whole element collapses. Then, the damage variable can be interpreted as the ratio of the damaged area, A_d, with respect to the area A^e associated with the local discretization. In the context of finite elements, the area A^e represents the area of the element (or that of an integration point). Using the linear softening law represented in Fig. 1, this ratio is a function of the energy dissipated during the damage process Ξ, and of the critical energy release rate G_c. The damage variable can be expressed as

$$\bar{d}=\frac{A_d}{A^e}=\frac{\Xi}{G_c} \tag{2}$$

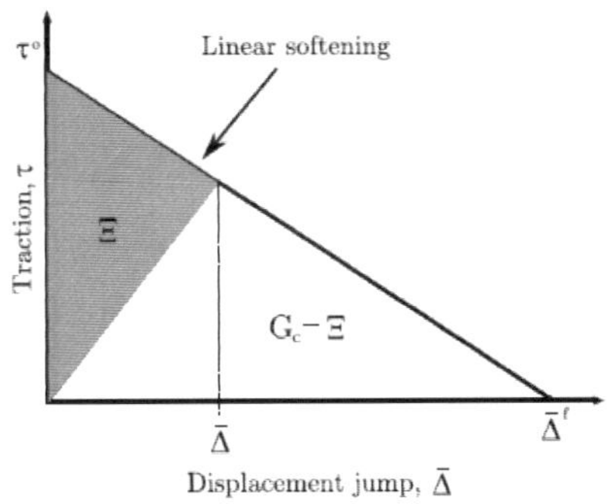

Figure 1. Linear softening law for a cohesive zone model approach.

2.2. Damage evolution criteria

Damage evolution criteria for none cohesive zones apply Hashin's failure criteria in which the failure modes are as follows.

1) Tensile fibre failure for $\sigma_{11}\geq 0$

$$\left(\frac{\sigma_{11}}{X_T}\right)^2+\frac{\sigma_{12}^2+\sigma_{13}^2}{S_{12}^2}=\begin{cases}\geq 1 & \text{failure}\\ <1 & \text{no failure}\end{cases} \tag{3}$$

2) Compressive fiber failure for $\sigma_{11} < 0$

$$\left(\frac{\sigma_{11}}{X_C}\right)^2=\begin{cases}\geq 1 & \text{failure}\\ <1 & \text{no failure}\end{cases} \tag{4}$$

3) Tensile matrix failure for $\sigma22 + \sigma33 > 0$

$$\frac{\left(\sigma_{22}+\sigma_{33}\right)^2}{Y_T^2}+\frac{\sigma_{23}^2-\sigma_{22}\sigma_{33}}{S_{23}^2}+\frac{\sigma_{12}^2+\sigma_{13}^2}{S_{12}^2}=\begin{cases}\geq 1 & \text{failure}\\ <1 & \text{no failure}\end{cases} \tag{5}$$

4) Compressive matrix failure for $\sigma_{22}+\sigma_{33} < 0$

$$\left[\left(\frac{Y_C}{2S_{23}}\right)^2-1\right]\left(\frac{\sigma_{22}+\sigma_{33}}{Y_C}\right)+\frac{\left(\sigma_{22}+\sigma_{33}\right)^2}{4S_{23}^2}+\frac{\sigma_{23}^2-\sigma_{22}\sigma_{33}}{S_{23}^2}+\frac{\sigma_{12}^2+\sigma_{13}^2}{S_{12}^2}=\begin{cases}\geq 1 & \text{failure}\\ <1 & \text{no failure}\end{cases} \tag{6}$$

Where, σ_{ij} denote the stress components and the tensile and compressive allowable strengths for lamina are denoted by subscripts T and C, respectively. X_T, Y_T, Z_T denotes the allowable tensile strengths in three respective material directions. S_{12}, S_{13} and S_{23} denote allowable shear strengths in the respective principal material directions.

IOP Conf. Series: Materials Science and Engineering **326** (2018) 012004 doi:10.1088/1757-899X/326/1/012004

2.3. Multicontinuum theory

Fe-safe is a robust and computationally efficient tool for predicting fatigue life in composite structures. By utilizing an Multicontinuum theory (MCT), fe-safe helps to get access to constituent level fatigue behavior in a full-scale structural analysis, which is difficult for conventional structural analysis, see figure 2. The fundamental premise underlying continuum mechanics is that any physical quantity of interest can be evaluated at a material point by averaging the quantity over a representative volume element that surrounds the point of interest. The size of this representative volume element (RVE) must be very small compared to the overall physical dimensions of the material body, yet large enough to provide an accurate statistical representation of the quantity within the body's material microstructure. For typical unidirectional, fiber reinforced composites, the fibers tend to have a somewhat random spacing within the matrix material. Therefore, the RVE used to characterize a material point must be large enough to contain numerous fibers to provide an accurate statistical representation of any quantity averaged over the RVE. However, for the sake of computational expediency, it is common practice to assume a uniform fiber distribution that represents the actual random fiber distribution in some statistically meaningful sense. This assumption of uniform fiber spacing permits the RVE to economically be represented by a single unit cell with periodic boundary conditions. [14].

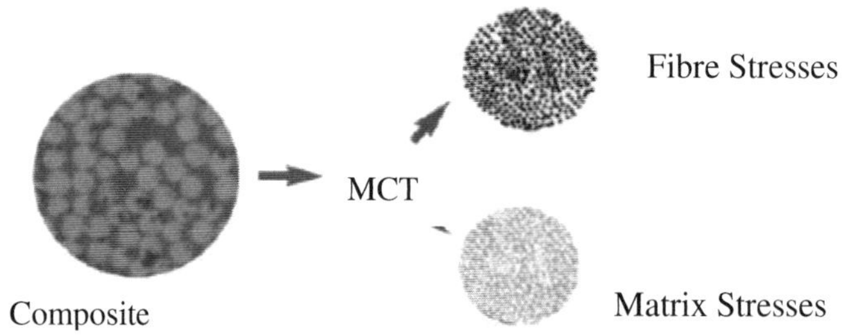

Figure 2. Composite lamina as a multicontinuum.

2.4. Fatigue failure criterion

The current constituent level failure criteria are based on revised failure criteria originally proposed by Mayes [15]. He developed a simple quadratic stress-based failure criteria to predict constituent failure within a composite lamina.

$$^{\pm}A_1^f\left(I_2^f\right)^2 + A_4^f I_4^f = 1 \tag{7}$$

$$^{\pm}A_1^m\left(I_1^m\right)^2 - {}^{\pm}A_2^m\left(I_2^m\right)^2 + A_3^m I_3^m + A_4^m I_4^m - {}^{\pm}A_5^m I_1^m I_2^m = 1 \tag{8}$$

Where

$$\begin{aligned} I_1 &= \sigma_{11} \\ I_2 &= \sigma_{22} + \sigma_{33} \\ I_3 &= \left(\sigma_{22}\right)^2 + \left(\sigma_{33}\right)^2 + 2\left(\sigma_{23}\right)^2 \\ I_4 &= \left(\sigma_{12}\right)^2 + \left(\sigma_{13}\right)^2 \end{aligned} \tag{9}$$

In above equation, the quantities I_j^{α} (j = 1,2,3,4,5) are transversely isotropic invariants of the matrix average stress state. $\alpha = f$ for fiber, $\alpha = m$ for matrix. The coefficients A_j^m and A_2^f , leading the invariants, are constituent failure parameters, generally derived from experimentally determined composite ultimate strength data through correlation with the MCT decomposition. The ± signs in these equations indicate a dependence of the parameter on tensile versus compressive stress. Details on the computation of these parameters for the analyses presented here-in may be found in Kenik [16].

3. Finite element model

Consider a half piece of a composite(T800H-2500-RT) single lap joint with a bolt hole in the center, see figure3. L=240mm， W=36mm, H=3mm. Radius of the hole is 3mm.The layer construction of the plate from top to bottom is $[45°/0°/-45°/0°/90°/0°/45°/0°/-45°/0°]_s$. Construct the initial delamination damage around the bolted hole by cutting off two elliptical defects, see figure 4. The two defects are located respectively at the bottom of the first layer and the second layer with a depth of 0.01mm. Their major radius and minor radius are 6mm and 4.5mm. As delamination damage degrades material property of the top two laminar layer, mainly on adhesive layers, two 0.01mm thin geometry layers with the same level of the elliptical defects were spilt from the plate. This two geometry zones were assigned as cohesive zone. Thus this two thin layers were applied with mesh category of COH3D8. The rest zones of the plate were applied with mesh category of SC8R. The left side of the plate is clamped, the right side is exerted with a linear displacement loading which is 2mm. A dynamic explicit method was used to compute the field variables with the time period of 0.001s. Material properties are listed in table 1 and table 2.

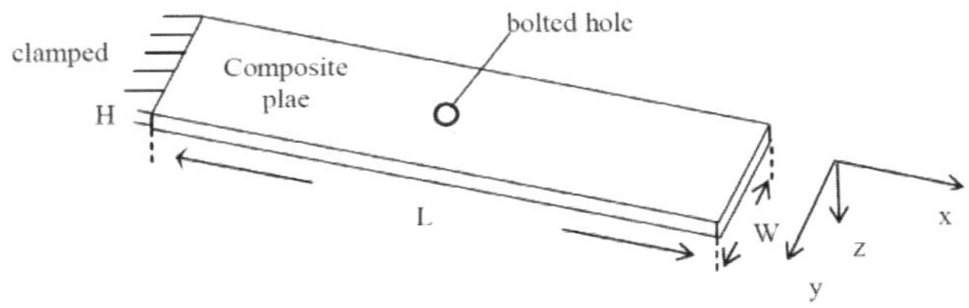

Figure 3. Composite plate with a bolted hole.

(a) delamination zone at top laminar layer. (b) delamination zones at 2nd laminar layer.

Figure 4. Shapes of delamination zones for composite plate with bolted hole.

Table 1. Material properties of composite plate with a bolted hole.

Material	T800H-2500-RT				
Elastic properties		Strength		Fracture toughness	
E_1(GPa)	153	X_t (MPa)	2537	G_{1C}^T (N/mm)	91.6
E_2=E_3 (GPa)	10.3	X_c (MPa)	1580	G_{1C}^C (N/mm)	79.9
$\nu_{12}=\nu_{13}$	0.3	Y_t (MPa)	82	G_{2C}^T (N/mm)	0.22
ν_{23}	0.4	Y_c (MPa)	236	G_{2C}^C (N/mm)	1.1
$G_{12}=G_{13}$ (GPa)	6	$S_{12}=S_{13}$ (MPa)	90	G_S(N/mm)	0.7
G_{23}(GPa)	3.7	S_{23} (MPa)	40		

Table 2. Properties of cohesive zones.

Properties of cohesive elements	Mode I	Mode II	ModeIII
Normalized elastic modulus (GPa)	120	43	43
Interlaminar strength (MPa)	30	80	80
Interlaminar fracture toughness (N/mm)	0.52	0.92	0.92

4. Results and discussion

A mesh convergence study was carried out to ensure that the results of FEA models are reliable, see table 3. As the results show, the max Mises stress varies little after the mesh number exceeding 10^4. In the subsequent work, a model with 23351 meshes elements is adopted, comprehensively considering the computing efficiency and accuracy.

Table 3. Max Mises stress for cases with different amount of meshes.

sample	Number of meshes	Max Mises stress
1	13104	2421.44
2	23351	2435.06
3	32023	2438.38

To implement fatigue analysis, acquiring initial stress or strain distribution is of vast important, especially for stress concentration region near the bolted hole. In this study, fatigue based on cyclic stress is researched. As is shown in figure 5, three cases with different element density attained pretty much the same stress field results, and the highest-value point located at basically the same place. The reason why stress concentration areas are not uniformly distributed around the hole is that the arrange style of every ply is different making the property of material anisotropic. However, we can conclude that the adopted mesh strategy achieved stable solutions.

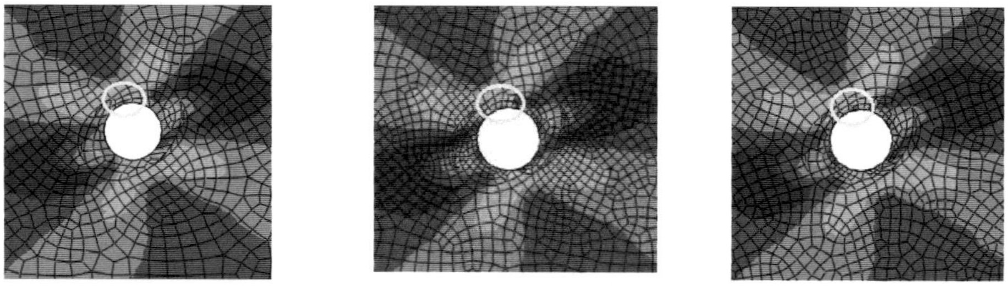

(a) case a with 13104 mesh elements. (b) case b with 23351 mesh elements. (c) case c with 32023 mesh elements.

Figure 5. Top layer stress distribution for regions of interested.

Tensile fiber failure and tensile matrix failure of the most dangerous element, as seen in the yellow circle of figure 5, is investigated using normalized damage index with respect to the end displacement(or loading), see in figure 6. Damage occurs when tensile damage index is over 0, material collapse when tensile damage index reaches 1. It is observed that the delamination damage of composite structure evolutes in an accumulative way with increasing of end loading, and the matrix is much earlier to lose bearing capacity than the fiber. What's more, the plate with delamination defects is much easier and much earlier to get failed than the perfect plate under the same loading conditions. The results are in accord well with physical fact.

Using Fe-safe to import results of above analysis as the basic data can generate S-N curve for the composite plate in study. Then define the material parameters and loading history, we can calculate the composite plate's fatigue life. In this case, loading history is set as $[n,0,-n]_m$, where n is loading amplitude, [n,0,-n] is a basic period, m is cycle number.

Figure 7 draws that fatigue life of the composite plate with a bolt-hole drop rapidly with respect to the load amplitude increase, furthermore, an increase of 67% load amplitude leads to a decrease of 99.5% fatigue life. This demonstrates that fatigue life of composite bolted joints with bolt-hole delamination is highly sensitive to external loads. It also reveals the value and importance of studying the fatigue life for composite structures with defects.

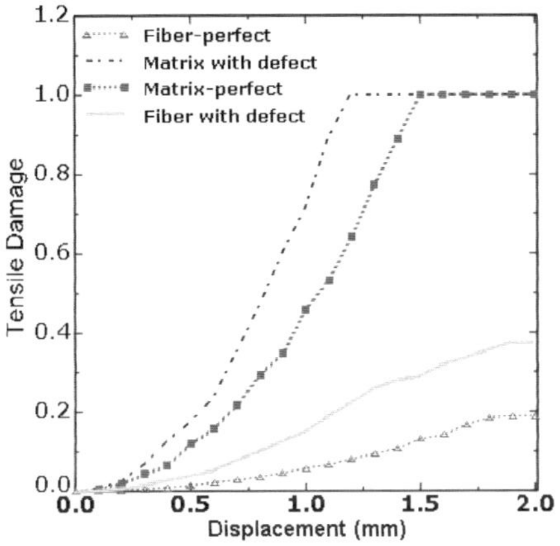

Figure 6. Fiber/matrix damage evolution in defect/perfect structures.

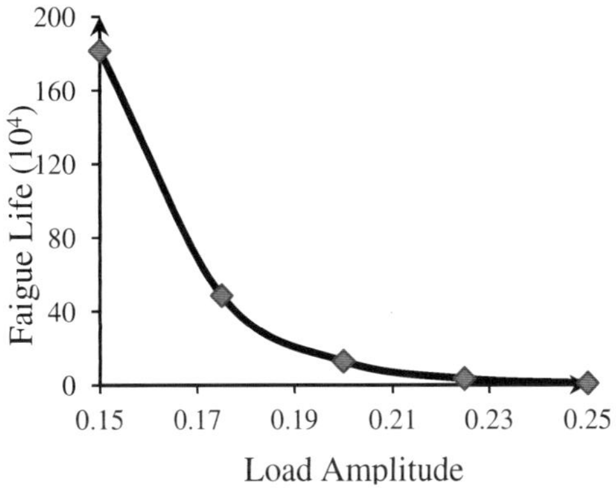

Figure 7. Load amplitude vs fatigue life.

Take the case with load history of $[2.5,0,-2.5]_m$ for example to analyze fatigue results. Figure 8 illustrates fatigue life of every element on the top layer of the zone around the hole. Compared with figure 5, we can see that stress concentration areas possess the lowest cycle number, while low stress area is able to cycle millions of times remaining good service. This phenomenon results from the fact that high stress areas take possession of higher probability to accumulate irreversible plastic deformation which leads to fatigue failure. It is worth noting that this result only explained local composite fatigue, since most of the rest areas still hold load-bearing capacity. If the full-scale structure's fatigue performance is needed, one solution is to modify the above local fatigue region's property and recalculate the new model's fatigue life until the overall fatigue takes place.

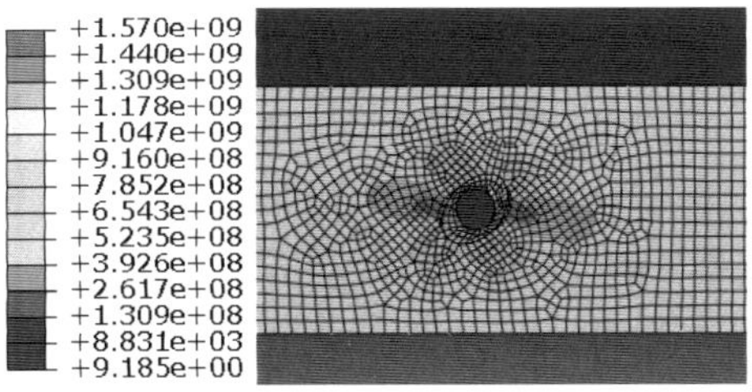

Figure 8. Fatigue life of top layer for the zone around the hole.

5. Conclusions

Bolt-hole delamination evolution of composite bolted joints is briefly investigated using CZM combined with Hashin's failure criteria, and the influence of delamination damage to composite material is discussed in virtue of commercial codes fe-safe. This numerical approach for studying fatigue life of composite structures with local delamination damage is quite suitable for engineering application. The below are the conclusions derived from the present study carried out on the composite bolted joints with bolt-hole delamination under cyclic symmetric tension-compression load.

(1) CZM is an effective way to simulate bolt-hole delamination evolution for composite structures, and slight degradation of material properties for local region with geometric singularity obviously accelerates local failure.

(2) Fatigue life of bolt-hole delamination area drastically decrease with the increase of end load, this is particularly evident at high stress area. For the composite plate in this study, an increase of 67% load amplitude leads to a decrease of 99.5% local fatigue life.

Acknowledgments

This paper was supported by the National Basic Research Program of China (Grant No. 2014CB046506), the National Natural Science Foundation of China (Grant No. 51105056) and the Fundamental Research Funds for the Central Universities (Grant No. DUT15QY16).

References

[1] Halpin J C. 1972 *J International Journal of Fracture* **8(4)** 465-468
[2] Chou P C, Croman R. 1978 *J. Journal of Composite Material* **12(2)**177-194
[3] Yang J N, Lee L J, Sheu D Y. 1992 *J. Composite Structures* **21(2)**91-100
[4] Feng G H, Yang Y Q, Luo X, et al. 2015 *J Composites Part B: Engineering***68** 336-342
[5] Capela C, Ferreira J A M, Febra T, et al. 2015 *J International Journal of Fatigue***70** 216-222
[6] Chaboche J L. 1998 *Journal of applied mechanics* **55(1)** 59-64
[7] Van Paepegem W, Degrieck J. 2005 *J Computers & structures* **83(23)** 1930-1942
[8] Obert E, Daghia F, Ladeveze P, et al. Micro and meso modeling of woven composites: Transverse cracking kinetics and homogenization J. Composite Structures, 2014, 117: 212-221
[9] Abdelal G F, Caceres A, Barbero E J. 2002 *J* **56(4)**413-422
[10] Daghia F, Ladevèze P. 2012 *JComposite structures* **94(12)**3644-3653
[11] Van Paepegem W, Degrieck J, De Baets P. 2001 *J Composites Part B: Engineering***32(7)** 575-588
[12] Shokrieh M M, Rafiee R.2006*J Composite Structures* **74(3)** 332-342.
[13] Shokrieh M M, Lessard L B. 2000 *J Journal of Composite Materials* **34(13)**: 1056-1080
[14] Mayes J S, Hansen A C. 2004 *J Composites Science and Technology* **64(3)**379-394.
[15] Mayes J S, Hansen A C. 2004 *J Composites Science and Technology* **64(3)** 517-527.
[16] Kenik D J M 2009 *Advanced techniques for constituent-based progressive failure analysis of composite structures* (Wyoming: University of Wyoming)

Design and Analysis of Wind Turbine Rotors Using Hinged Structures and Rods

Hongya Lu[1], Pan Zeng[2,*] and Liping Lei[2]

[1]Beijing Institute of Mechanical Equipment, No.50 Yongding Road, Haidian District, Beijing 100039, China
[2]Key Laboratory for Advanced Material Processing Technology of MOE, Department of Mechanical Engineering, Tsinghua University, Beijing 100084, China

*zengp@mail.tsinghua.edu.cn

Abstract. Light weight and high stiffness are key design factors in ensuring cost effectiveness and reliability of wind turbines, especially for the inboard region of the rotor blades. In this study, several novel designs were developed to improve the mechanical performance of the rotor. Experiments were performed on an isolated blade incorporating the new features of a hinged structure and rods. The results validated the effectiveness of these features at alleviating the root-bending moment of the blade under varying wind loads and enhancing the stiffness of the blade. A numerical investigation was carried out to further examine the bending moment distribution, shear and axial force, and rod tension of these novel rotor designs under uniform loads. Longitudinal geometrical variations of the blade were considered in the model. Results showed that two designs realized a favorable bending moment distribution and improved the modal frequencies of the edgewise modes: bisymmetrical rods on a single-hinged structure and interveined symmetrical rods on a cantilevered structure. However, these designs have different deformation mechanisms. In addition, the first group of edgewise modal frequencies of these two designs were improved compared with the traditional rotor design. Their potential values in the application to the design of a lightweight, high-stiffness, and reliable wind turbine rotor were discussed.

1. Introduction

The increasing size of wind turbine rotors has posed new demands for wind turbine blade designs [1] in order to ensure both a light weight and high stiffness and a low cost and high reliability. [2] The loads on the blade are the key factor that determine the structural design.[3] In-plane loads including gravity, the centrifugal force and the aerodynamic load for the large blades increase enormously and become not negligible.[1,4]

There have been several studies on realizing lightweight and stiff wind turbine rotors by using advanced materials. For instance, carbon fiber reinforced composite laminates have been used to enhance the stiffness and mass ratios of the blade.[5,6] However, the high cost of carbon fiber material and the difficulty with designing a gradual transition between carbon fiber and ordinary glass fiber have limited the application of this approach. Other designs include adopting a new airfoil to meet the stiffness requirement. The flat-back foil with thick or blunt trailing edge [7, 8] have been utilized at the blade root to enhance the cross-sectional area and bending stiffness. SNL 100-03 includes a flat-back trailing-edge foil to realize a lower moment at the blade root and adopts a slender foil combined with the optimized structural and material configuration to realize a 16% reduction in weight compared with the SNL 100-02 blade. The present authors [9] aimed to reduce the moment of the blade at the root by introducing a rod design. A combined numerical and experimental investigation into the structure indicated that the weight of an NREL 5 MW blade is reduced by approximately one-third.

A novel design of the blade root with a small bending load and high stiffness is necessary to obtain a lightweight and reliable design of the rotor. The conventional rotor has a cantilevered beam structure, which does not reduce the loads at the root. Gourieres and Shi [10] presented a design that releases a degree of freedom (DOF) of the blade and supporting structure to reduce the load. The present authors previously considered the concept of releasing a DOF[9] to develop a specific rotor design.

This paper presents a more elaborate investigation of several rotor designs that considers both the load-bearing characteristics and stiffness. The design features of a hinged structure and tension rods were considered. Experiments and the finite element method were utilized to reveal the principles behind the mechanical behavior of these designs. The differences between these designs were considered. Two schemes were selected owing to their good performance, and the factors that influenced their enhanced stiffness were compared.

2. Novelwind turbine rotor design schemes

Scheme 1 (Figure 1) shows the structure of a conventional wind turbine rotor: three blades are fixed at the hub to form three cantilevered beams. A hinged structure and tension rods were used to design seven novel structures with the aim of improving the moment distribution on the blade by changing the configuration of the cantilevered beam.

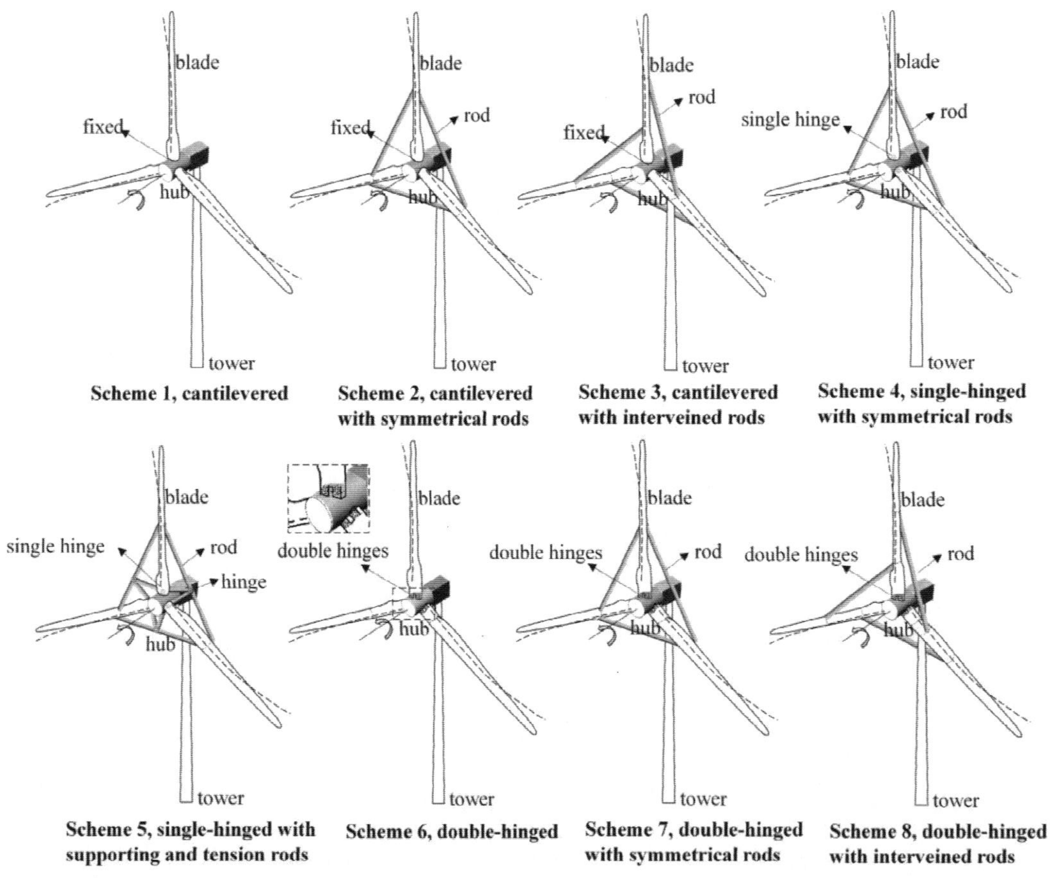

Figure 1. Illustrations of novel wind turbine rotors featuring a hinged structure and rods.

In scheme 2, three rods are mounted symmetrically around the hub between the blades. The three blades are still fixed to the hub, and each rod is located at the same position on the blade. Scheme 3 reorganizes the rods between the blades; the two ends of each rod are located at different positions on the blades (i.e., interveined). In scheme 4, the three blades are connected with the hub by a hinged structure. Because a DOF of the blade in the rotor rotating plane is released, three central symmetrical rods are introduced to support the structure. This structure is actually unstable. When one of the blades deflects slightly, the whole rotor rotates around the hub to arrive at a new balanced state. Scheme 5 upgrades scheme 4 to connect another three rods, called supporting rods, with the tension rods by hinged joints. The present authors [9] previously investigated scheme 5 with regard to the bending moment and tension on the rods under a concentrated force. Scheme 6 adopts a double-hinged joint similar to that applied in aircraft design [11] to connect the blades with the hub. This guarantees the stability of the structure without rods. Schemes 7 and 8 respectively introduce central and interveined symmetrical rods to scheme 6.

The hinged structure and rod arrangement style change the boundary conditions and stiffness distribution of the blade. The natural frequency and load distribution may differ from that of the conventional cantilevered blade structure. In addition, the tradeoff between a light weight and high stiffness may unbalance a design and make it unable to satisfy both requirements. Therefore, the influence of the hinged structure and rods on the mechanical behavior was investigated.All of the above designs were examined in terms of the rotating plane; thus, out-of-plane loads were not considered.

3. Experimental tests

First, the load-bearing characteristics with the hinged structure and rods were investigated. The method, used by H. Lu, was utilized to record the force and moment at the blade root. [9] Two kinds of structures were tested under wind velocities of 1–12 m/s: a hinged isolated blade with supporting rods and a conventional cantilevered blade, as shown in Figure 2.The truss root of the novel blade (left one) made it convenient to mount the rods. The rods were connected with the blade at the location of 15% span along the blade length. To prevent the vibrations of the wind tunnel platform affecting the results, the tested structures and sensors were all placed on a plate directly connected with the ground and separate from the bottom of the wind tunnel. When the chord of the blade tip was parallel to the wind direction, the angle of attack (AOA) was set as 0° . The experiments were conducted at AOAs of 15° and 20° for at least 30 s at each wind velocity.

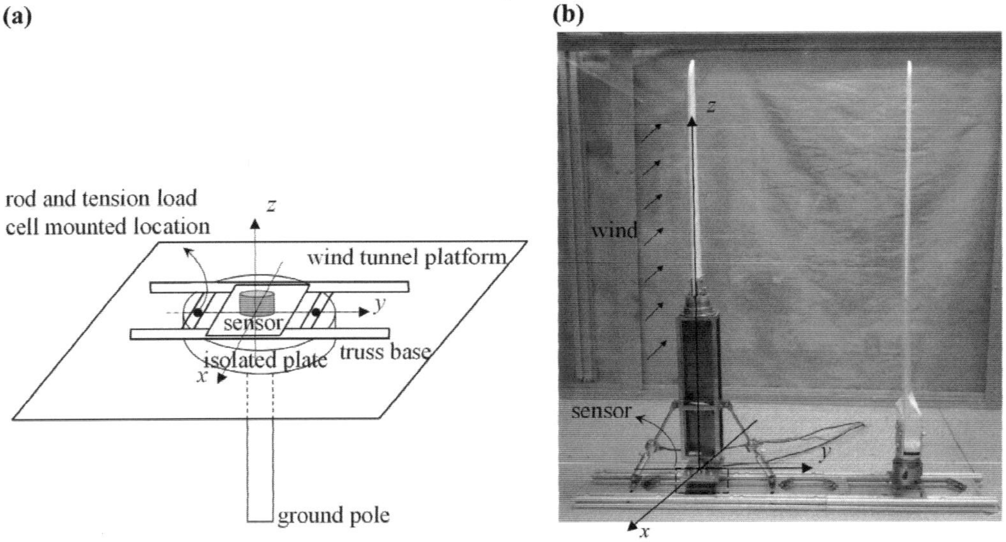

Figure 2. Wind tunnel platform and test details: (a) experimental setup; (b) two test structures.

The blade with hinged rods had load-bearing characteristics distinct from those of the cantilevered blade. As Figure 3(a) shown, at wind velocities of 1–12 m/s, the bending moment M_x at the blade root stayed approximately 0. The overall lift F_y tended to increase with the wind velocity. The hinged structure clearly reduced M_x, even at a wind velocity of 12 m/s. The tension in the rods demonstrated that they helped support the loads acting on the blade (Figure 3(b)). One rod bore the tensile load, while the other rod bore the compressive load; they were distributed almost symmetrically. The experimental results at AOAs of 15° and 20° demonstrated similar performances. The rod and a part of the blade below the location where the hinged rods meet the blade becomes an "amplified" hub, which makes it possible to transfer part of the load on the blade to the rods. In this manner, the bending moment on the inboard blade is reduced, which decreases the material needed for manufacture.

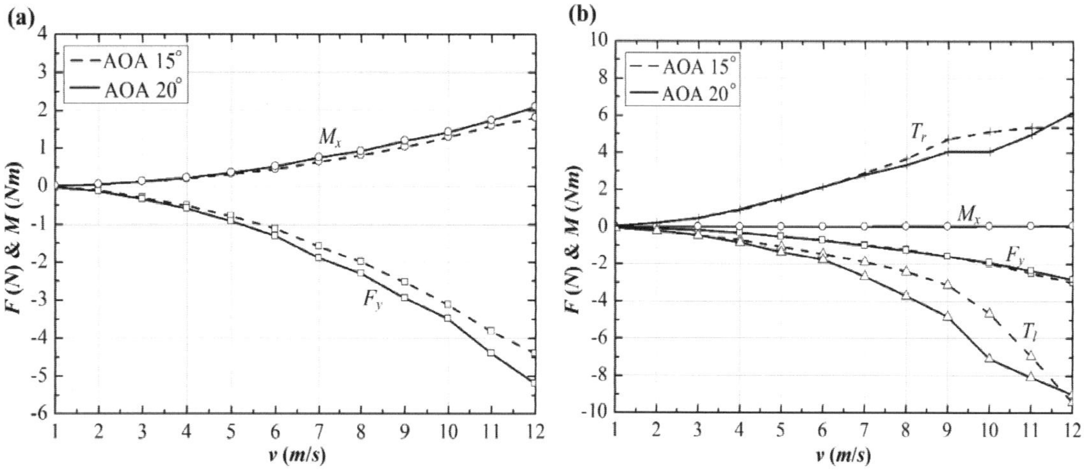

Figure 3. Experimental results,(a) the cantilevered blade at wind velocities of 1–12 m/s wind velocity: lift and bending moment; (b) the blade with rods at wind velocities of 3–12 m/s.

4. Numerical simulation of wind turbine rotor

A finite element model was established to study the mechanical behavior of rotors incorporating three blades, a hinged structure and tension rods (Figure 1). The bending moments, shear forces, axial forces, tension on rods and edgewise frequency were evaluated. ANSYS 14.5 were used for the models, which had exactly the same blades and rods, identical geometries and material properties under identical uniform loads.

4.1. Finite element model of the wind turbine rotor

Figure 4 shows the coordinate system. The rotor was located on the XOY plane. Uniform loads were distributed on perpendicular to each blade. Because the "box" of the main spar bears most of the loads on the blade, the blade was assumed to be segmented, and each segment had the same cross-section of a rectangular hollow tube. For both simplicity and effectiveness, the beam 188 element was used to describe the blade. The size of the element cross-section was derived by scaling the area and stiffness distribution of a NREL 5 MW blade. [12] Then, all available parameters were normalized by the data of the root cross-section. The obtained normalized edgewise stiffness distribution was compared with that of the NREL blade (Figure 4 (b)). The blade was scaled to a length of 2 m and divided into 17 elements. The nodes were numbered 1–18 from the root to the blade tip parallel to the *Y*-axis. The effective elastic modulus varies along the blade and the density was assumed to be 1800 kg/m^3. All of the rods were modeled as link 180 elements made of aluminum 6061. The double-hinged joint comprised an aluminum 6061 herringbone bracket with an angle between the two struts of 120°. When the blade was connected with the hub by the hinged joint, a corresponding rotation DOF was released. Otherwise, the blades were fixed at the root nodes. For all of the central symmetrical rods, the two rod ends were located at node 6 on the blades, which is 32.85% span along the blade length. For all of the interveined symmetrical rods, the two rod ends were located at nodes 4 and 8 on the blade, which are 16.59% and 45.86% span along the blade length separately. Static structural analysis was then conducted.

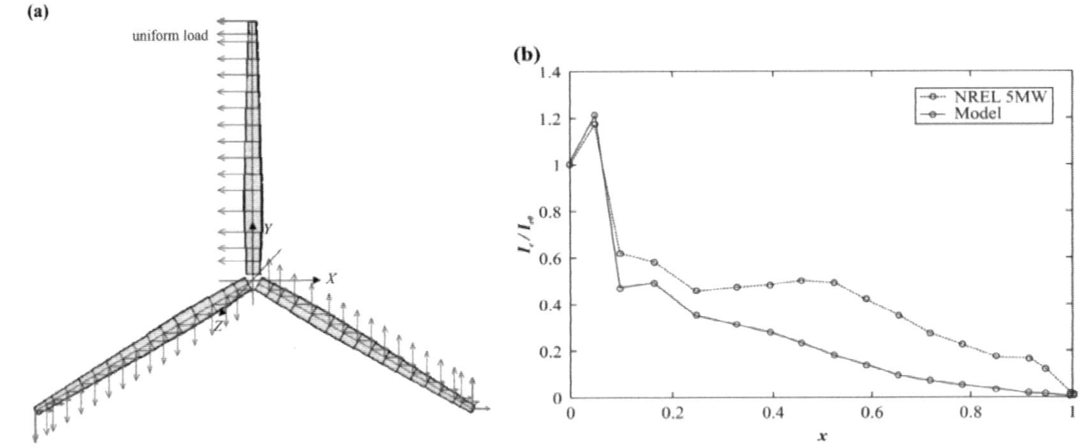

Figure 4. Finite element model of the rotor: (a) beam shape along the longitudinal axis and applied loads; (b) normalized edgewise stiffness distribution along the blade.

4.2. Static load-bearing characteristics

Figure 5(a) and (b) illustrate the in-plane bending moment and shear force when uniform loads act on a rotor with cantilevered blades. The shear force on the blade was linear, while the bending moment was quadratic from the root to tip. In scheme 2, the bending moment and shear force distribution were the same as in scheme 1, and the tension on the rods was zero. This can be explained by taking the blade along the Y-axis as an example. The two rods cause equal forces that are opposite in direction on the blade, so they cancel. In the steady state, the three rods bore no loads. Thus, the configuration of central symmetrical rods could not improve the moment distribution of the blade and had no effect from a load-bearing perspective. Scheme 3 changed the shear force distribution and bending moment distribution compared to that of the cantilevered blade (Figure 5(c) and (d)). The bending moment was decreased where the rod was mounted to the blade root. The bending moment reached a local minimum when the rod was close to the root. The maximum bending moment was still at the root but smaller.

Although the three rods had identical tensions, the ends of each rod were connected at different locations on the blade. This applied an axial force on the blade. The segment closest to the root had the largest axial force. The axial force helped reduce the bending moment below the rod-mounting location. Thus, the interveined symmetrical rods changed the blade deformation type. In this case, the blade tip experienced pure bending deformation, while the bottom part experienced both bending and compression deformations. Therefore, scheme 3 requires ensuring the axial compressive strength of the material. The hinged structure reduced the bending moment at the blade root to zero in the experiment.

Figure 6 in the Appendix shows the performance of scheme 5. The maximum bending moment was at the tension rod location and was the same as that of scheme 1, cantilevered. The tension in the supporting and tension rods acted together to share the load and reduce the bending moment on the inboard part of the blade. The influence of the rod locations on the moment distribution is discussed in detail in our previous report. [9]

Figure 7 in the Appendix illustrates the load distribution in scheme 6. The double-hinged structure changed the continuity of the shear force. The bending moment direction in the hinged structure became opposite that in the blade. Because the blade was identical with the previous cases, the maximum bending moment was still at the blade root. The bending moment and shear force in the herringbone bracket were symmetrically distributed. Because the bending moment at the joint of the blade and hinged structure is large, this part needs a detailed design to prevent stress concentration.

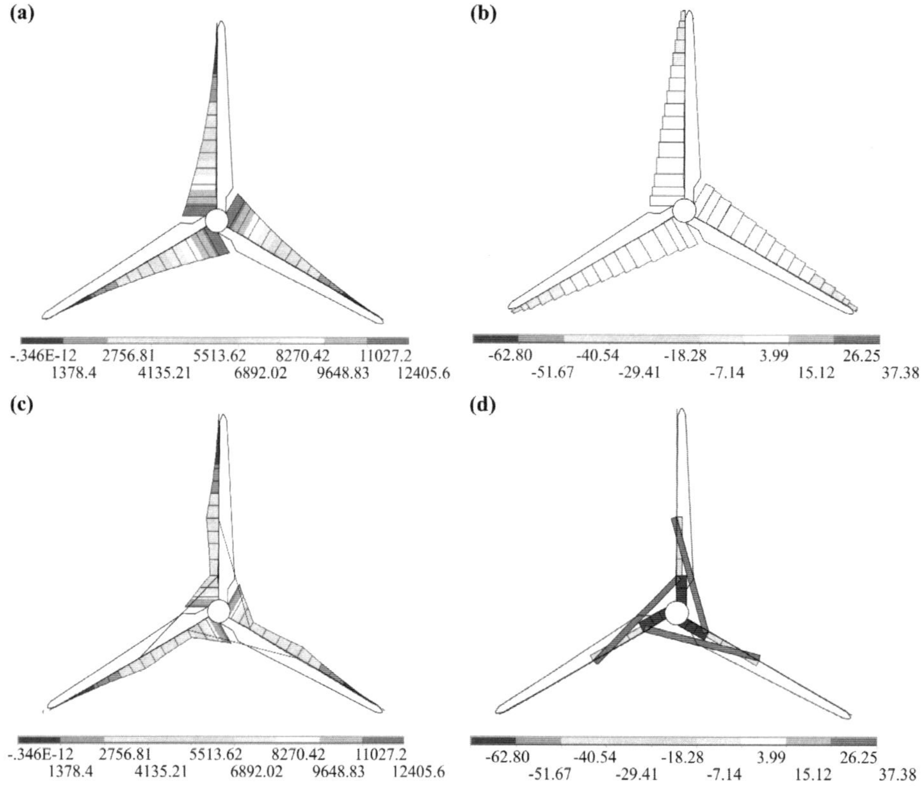

Figure 5. In-plane loads distribution, (a) scheme 1: bending moment (N-mm); (b) scheme 1: shear force (N); (c) scheme 3: bending moment (N-mm); (d) scheme 3: tension in rods and axial force in blades (N).

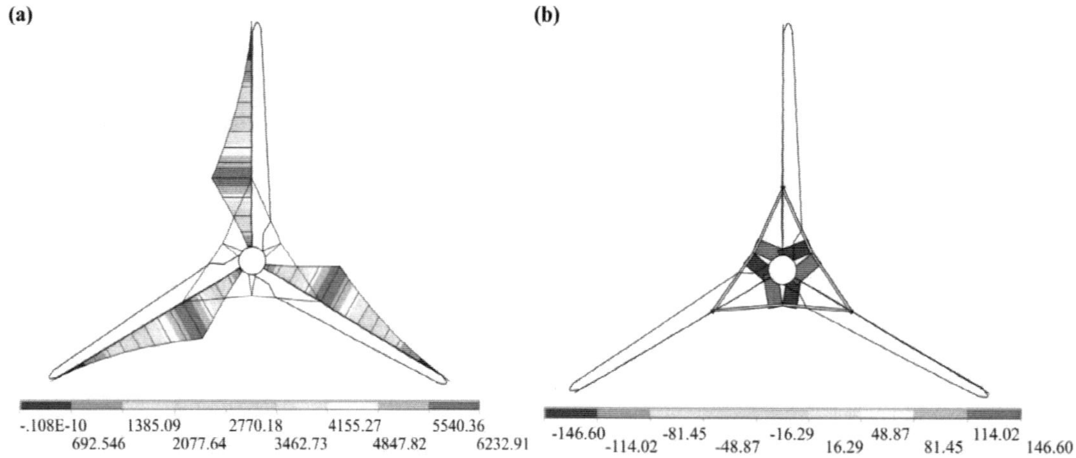

Figure 6. In-plane load distribution in scheme 5: (a) bending moment (N-mm); (b) tension in rods (N).

Figure 8 in the Appendix illustrates the influence of scheme 7. The bending moment kept the same distribution as scheme 6, and the rods bore no loads. The double-hinged structure was subjected to axial forces, where one side was tensile and the other was compressive. Scheme 8 combined the bending moment distribution and tension in rods in schemes 3 and 6 (Figure 9). The joint connecting the blade and double-hinged structure bore the largest bending moment. The local maxima bending moments occurred when the interveined rods were mounted at 16.59% span and 45.86% span along the blade length separately.

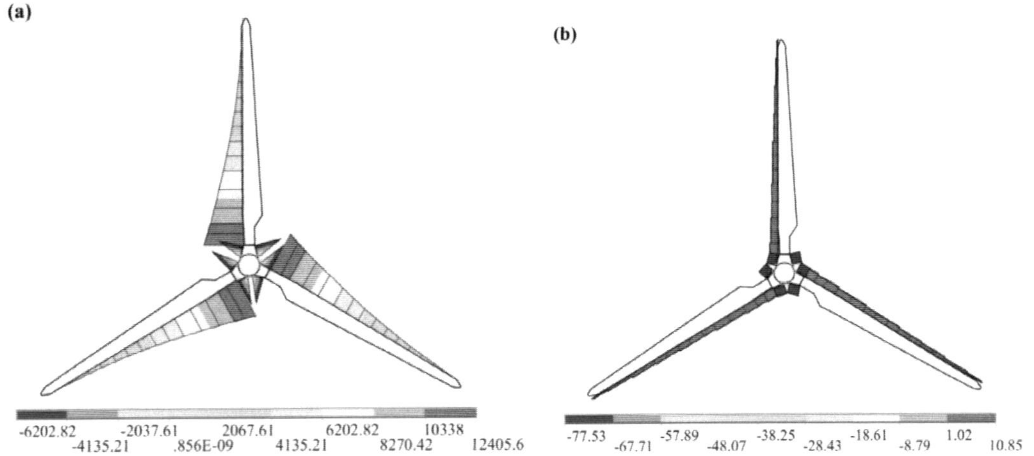

Figure 7. In-plane load distribution in scheme 6: (a) bending moment (N-mm); (b) shear force (N).

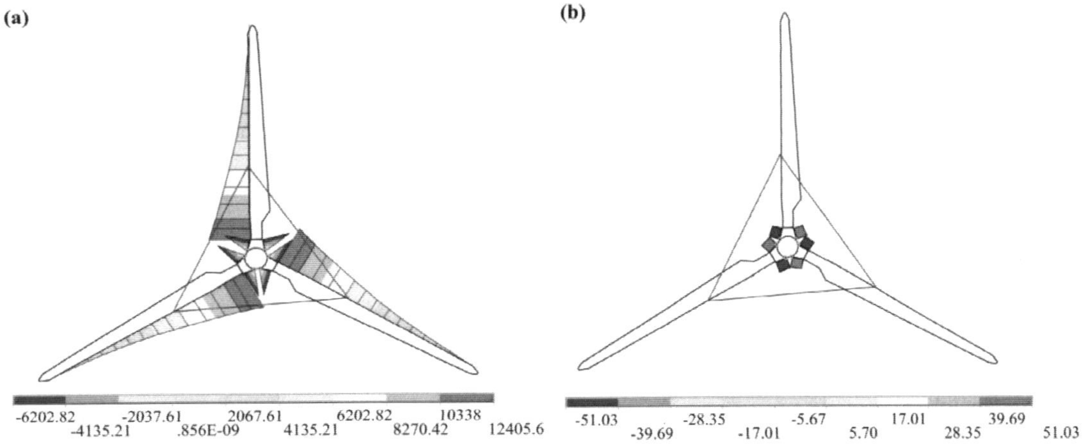

Figure 8. In-plane load distribution in scheme 7: (a) bending moment (N-mm); (b) tension in rods (N).

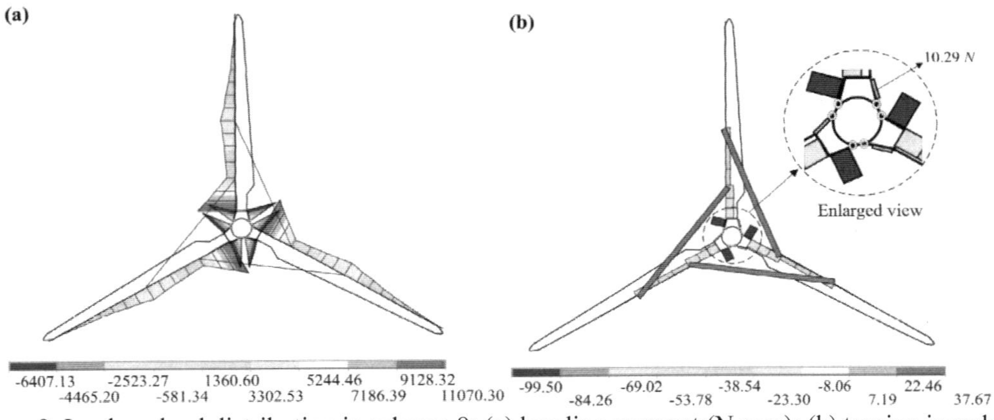

Figure 9. In-plane load distribution in scheme 8: (a) bending moment (N-mm); (b) tension in rods (N).

In scheme 8, the interveined symmetrical rods changed the axial force in the double-hinged structure. The load was greater in the brace bearing the compression deformation and was 99.50 N. The tensile load on the other side was 10.29 N. When schemes 3 and 8 were compared, the axial force below 45.86% span along the blade length in the double-hinged structure slightly exceeded that in the cantilevered blade. In scheme 8, the axial forces on the blade were 63.48 and 36.08 N. The tension in the three rods was identical at 37.67 N each.

The maximum bending moment and axial force of the above schemes were extracted (Figure 10). Scheme 5 achieved the largest reduction in bending moment and did not introduce an axial force. In addition, schemes 3 and 8, which both employed interveined symmetrical rods, decreased the maximum bending moment compared to the cantilevered blade because they introduced an axial force that bore part of the load.

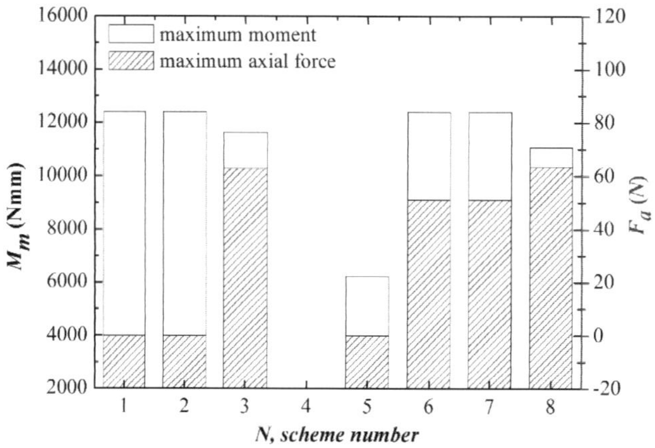

Figure 10. Maximum bending moment and axial forces of all schemes under a uniform load; *N* is the scheme number.

4.3. Modal analysis of the novel rotors

Modal analysis was performed based on the same structural models given above, and the first group of edgewise modes was extracted. Figure 11(a) shows the first group of edgewise modes in sequence. Figure 11(b) uses the central symmetrical rods in scheme 2 to represent the typical influence of the rods on the mode shapes. In the first group of edgewise modes, the three blades vibrated in sequence under an identical frequency. However, introducing the rods broke the isolation of the three blades. In the first edgewise mode, the three blades had the same deflections in the same direction in a centrosymmetric pattern. This was the collective mode of the rotating rotor. The flexibility of the in-between rods decreased the modal frequency of this mode. The other two mode shapes behaved like the whirling mode of the rotating rotor. When one blade reached the location of the largest deflection, the other blades arrived at different states but had the same frequency.

Other schemes with different rod configurations had similar mode shapes for the edgewise mode groups (Figure 11(b)). The modal frequency of the first edgewise mode group of the cantilevered rotor in scheme 1 is denoted as f_{e0}. Modal frequencies of the first edgewise mode group in the other schemes are denoted as f. All of the frequencies derived from the numerical simulation were normalized by f_{e0} (Figure 11(c)). Schemes 1 and 6, which had no rods, had an identical frequency in the first edgewise mode group. Once the rods were incorporated into the rotor, the three frequencies were grouped into two. The lower one was the first edgewise mode (i.e., collective mode), while the two identical higher frequencies were the whirling modes. The interveined symmetrical rods in schemes 3 and 8 increased all three frequencies of the first edgewise mode group. The central symmetrical rods only increased the two frequencies of the whirling modes but lowered the frequency of the collective mode. Scheme 5 improved all of the frequencies of the first group of edgewise modes. Therefore, schemes 3, 5 and 7 were effective at enhancing the rotor stiffness and ensuring reliability.

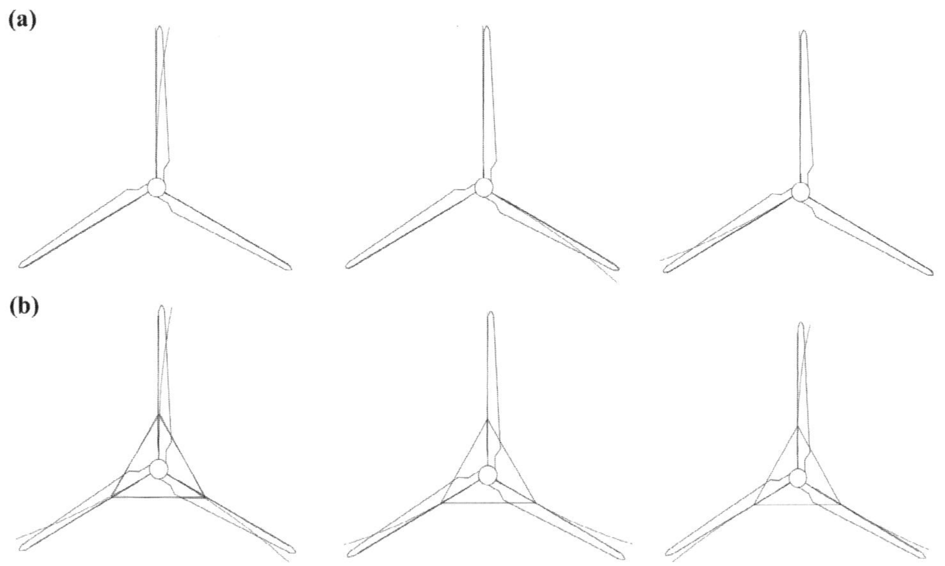

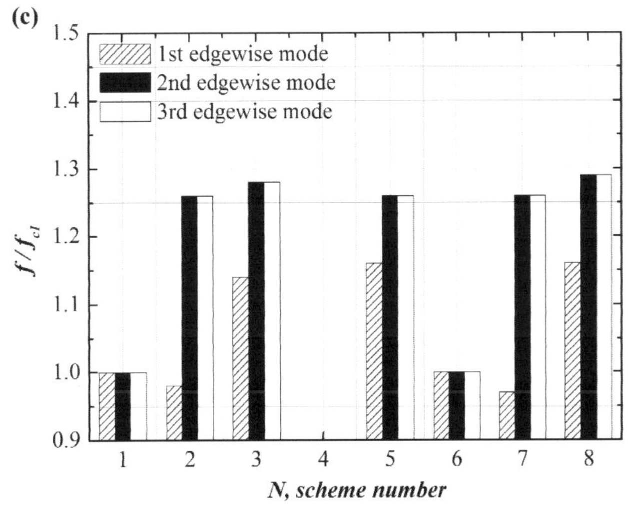

Figure 11. Influence of the rods on the mode shapes and frequencies of the rotor: (a) first group of edgewise modes in scheme 1; (b) first group of edgewise modes in scheme 2; (c) normalized modal frequency of the first group of edgewise modes of the rotors.

4.4. Potential application on the enhanced performance

In the study, the total weight of the blade was about 8.1 kg. For scheme 3, the weight of the added rods was 1.65kg, which accounted for 20% of the blade weight. If the total amount of material kept constant, it meaned that the weight of each blade was reduced to 80% of the original blade. Implementing this process by reducing the size of the spar, and repeating the numerical simulation, the mechanical performance of scheme 3 was obtained. The maximum moment at the root was 94.2% of the root bending moment on the cantilevered blades in scheme 1. Compared with scheme 1, the moment along the blades was smaller. The tension on the rods were 35.10 N.

Furthermore, the two schemes were compared regarding the natural frequencies. In scheme 3 with the same rotor weight as the scheme 1, the first group of edgewise frequencies were 45.675, 50.537

and 50.537 Hz. While in scheme 1, the first group of edgewise frequencies were identical, which were 42.505, 42.505 and 42.505 Hz. The comparison between these two schemes proved that the edgewise modal frequencies can be improved and the moments can be reduced by adopting the interveined rods in scheme 3.

The similar manipulation was operated on scheme 5 to keep the same rotor weight as scheme 1. The obtained moments were obviously smaller than that of scheme 1. The first group of edgewise frequencies were 42.339, 47.755 and 47.755 Hz. In the first edgewise modes group, the collective modal frequency kept almost the same while the whirling modal frequencies were improved.

According to the above analysis, when scheme 3 and scheme 5 kept the same weight with scheme 1, the blades endured less loads and became stiffer. Thus, utilizing schemes 3 and 5 to design a wind turbine rotor can reduce the material requirements.

5. Conclusions

In this study, a series of novel wind turbine rotors was designed featuring a hinged structure and tension rods. The mechanical behavior of these designs, including the static load-bearing performance and modal frequency, were investigated experimentally and numerically. Several conclusions were drawn.

1) The experiments in the wind tunnel platform demonstrated that the hinged structure can relieve the bending moment at the blade root under different wind velocities.
2) The ANSYS simulation illustrated the load-bearing characteristics of the design schemes, which helped in selecting designs with a relatively small bending moment and large stiffness: bisymmetrical rods on a single hinged structure (scheme 5) and interveined symmetrical rods on a cantilevered structure (scheme 3).

Compared with the conventional cantilevered structure, utilizing schemes 3 and 5 to design a wind turbine rotor can reduce the material requirements while ensuring stiffness and guarantee performance reliability. Since the novel designs were located close to the root of the blades, their influences on the aerodynamics and energy production can be limited with optimized design. More study should be done to make a thorough understanding of the aerodynamic performance of scheme 3 and 5.

Acknowledgments

The study was supported by the National Natural Science Foundation of China (No. 51575296). The authors gratefully acknowledge this support.

References

[1] D. T. Griffith and T. D. Ashwill, The Sandia 100-meter all-glass baseline wind turbine blade: SNL100-00. Report, Sandia National Laboratories, Albuquerque, NM, Report No. SAND2011-377, 2011.
[2] A. Chehouri, R.Younes, A. Ilinca and J. Perron, Review of performance optimization techniques applied to wind turbines.Appl Energy 2015; 142: 361–388. DOI: 10.1016/j.apenergy.2014.12.043.
[3] International Electrotechnical Commission, IEC 61400-1: Wind turbines part 1: Design requirements. International Electrotechnical Commission. 2005.
[4] J. L. Tangler, The evolution of rotor and blade design. Report, National Renewable Energy Laboratory, Golden, CO, 2000.
[5] T. D. Ashwill, Materials and innovations for large blade structures: research opportunities in wind energy technology. In: 50th AIAA structures, structural dynamics, and materials conference, Palm Springs, CA, May 4–7, 2009.
[6] X. Lachenal, S. Daynes and P. M. Weaver, Wind Energy 2013; 16(2): 283–307. DOI: 10.1002/we.531.
[7] F. Grasso, J Aircr 2013; 50(3): 975–81. DOI: 10.2514/1.C032182.

IOP Conf. Series: Materials Science and Engineering **326** (2018) 012005 doi:10.1088/1757-899X/326/1/012005

[8] J. P. Baker, C. P. Van Dam and B. L. Gilbert, Flatback airfoil wind tunnel experiment. Report, Sandia National Laboratories, Albuquerque, New Mexico, 2008.
[9] H. Lu, P. Zeng, L. Lei, Y. Yang, Y. Xu and L. Qian, Energy Convers Manag 2014; 88: 535–44. DOI: 10.1016/j.enconman.2014.08.051.
[10] D.L. Gourieres and P. F. Shi, Wind power plants: theory and design, 1st ed (China Machine Press, Beijing, 1987). [Translation]
[11] Z. Wang, W. Yao, Aircraft Structural Design, 1st ed (National defense industry press, 2004).
[12] S. Butterfield, W. Musial and G. Scott. Definition of a 5-MW reference wind turbine for offshore system development. Report, National Renewable Energy Laboratory, Golden, CO, 2009.

IOP Conf. Series: Materials Science and Engineering **326** (2018) 012006 doi:10.1088/1757-899X/326/1/012006

Quantitative Evaluation of Delamination in Composites Using Lamb Waves

L Michalcová and R Hron

Strength of Structures Department, Materials, technologies and NDT Group, VZLU - Czech Aerospace Research Center, Beranových 130, 19905 Prague, Czech Republic

michalcova@vzlu.cz

Abstract. Ultrasonic guided wave monitoring has become very popular in the area of structural health monitoring (SHM) of aerospace structures. Any possible type of damage must be reliably assessed. The paper deals with delamination length determination in DCB specimens using Lamb waves. An analytical equation based on the velocity dependence on variable thickness is utilized. The group velocity of the fundamental antisymmetric A0 mode rapidly changes in a particular range of the frequency-thickness product. Using the same actuation frequency the propagation velocity is different for delaminated structure. Lamb wave based delamination lengths were compared to the visually determined lengths. The method of the wave velocity determination proved to be essential. More accurate results were achieved by tracking the maximum amplitude of A0 mode than the first signal arrival. These findings are considered as the basis for the damage evaluation of complex structures.

1. Introduction

Increased use of composite materials in aerospace structures require reliable methods for assessment of different types of possible damages. Non-destructive testing (NDT) methods are considered as a standard practice [1, 2, 3, 4], however, to increase the aircraft safety and availability and to decrease the maintenance costs the structural health monitoring (SHM) approach has been implemented in the process of the structural design and testing. Some NDT and SHM methods use similar physical principles, but the different approach makes SHM on-line, condition-based and predictive monitoring with sensors permanently attached to the structure [5].

There are many SHM techniques under detailed research as Acoustic Emission (AE) [6] Fiber Bragg Grating sensors (FBG) [7, 8], Electrical Resistance Tomography (ERT) [9], Ultrasonic Guided Wave testing (UGW) [5, 10, 11] and others, of which FBG and UGW play an important role with probably the highest technical readiness level (TRL). Certification of such methods starts of course on the baseline of the building block approach and require not only the assessment of reliability of the methods themselves, but also the sensors, measuring devices, algorithms etc. FBG sensors are particularly intended for the stress/strain measurements and are considered as a good alternative to the strain gauges (SG) especially due to their immunity to the electromagnetic interferences. UGW testing showed more capabilities especially due to their ability to localize damage. Tomographic imaging algorithms became very popular particularly for the impact damage assessment [12, 13].

IOP Conf. Series: Materials Science and Engineering **326** (2018) 012006 doi:10.1088/1757-899X/326/1/012006

Aircrafts are operated in such conditions which naturally lead to susceptibility to damage – either the fatigue cracks in metals or delaminations in composites. Delamination is the most common damage in composites and occurs as a result of impact damages, compression, tension or bending loadings. The main purpose of DCB test is to determine the fracture toughness in Mode I (G_{IC}) and delamination lengths are needed for G_{IC} calculations. This paper deals with determination of delamination lengths using Lamb waves (UGW) measurements. The literature on this topic shows a variety of approaches, which correlate Lamb wave signal features with delamination propagation either experimentally or numerically. Proposed methods incorporated for evaluation different signal features as correlation coefficients, magnitude changes, velocity/time of flight (ToF) changes etc [14, 15]. Another work is based on the numerical assesment of the time of flight of direct and turning Lamb mode during crack propagation [16]. In presented paper delamination growth of double cantilever beam specimens (DCB) using Lamb waves is investigated from the point of the possibility of the precise calculation of delamination length. Precise delamination length is determined using change in the time of flight of A0 mode and group velocity in the original beam and in the beam of half thickness. The paper deals with the question of tracking either the peak with maximum amplitude of the wavefront or the first arrival time of the wave. Finally the both approaches of delamination lengths calculation using Lamb waves are compared to the visually based values.

2. Materials and methods

2.1. Mechanical testing

Specimens were made of carbon fiber plain weave fabric with epoxy resin using a sequence of lamination $[45/0]_{4S}$. A polymer foil with the length of approximately 25 mm was inserted into the midplane of the 16-ply laminated panels as an artificial initial delamination. Specimens with dimensions of 250x25 mm and thickness of 3 mm were cut from the panels. DCB tests were performed according to the ASTM D5528 standard. Determination of the fracture tougness in Mode I (G_{IC}) is the main purpose of this test. Delamination lengths are needed for the G_{IC} calculation and usually the lengths are measured approximately to half of the specimen length. Nevertheless, for the objective of Lamb wave measurements 5 DCB specimens were loaded until the final fracture – until the splitting the specimen into 2 parts.

2.2. Lamb wave propagation

Lamb waves belong to the group of ultrasonic guided waves and propagate with very low attenuation in structures with comparable or greater wavelength than the structure thickness. In general, these are plate-like structures or shells acting as a waveguide. Unlike bulk waves guided waves show multimodal and dispersive character which may be considered either as an advantage or a disadvantage. Dependency of phase and group velocities on the frequency-thickness product is described by dispersion curves (Figure 1). Two fundamental modes symmetric S0 and anti-symmetric A0 exist within the whole frequency-thickness spectra. Above the cut-off frequency (or the cut-off frequency – thickness product) increases the number of propagating modes.

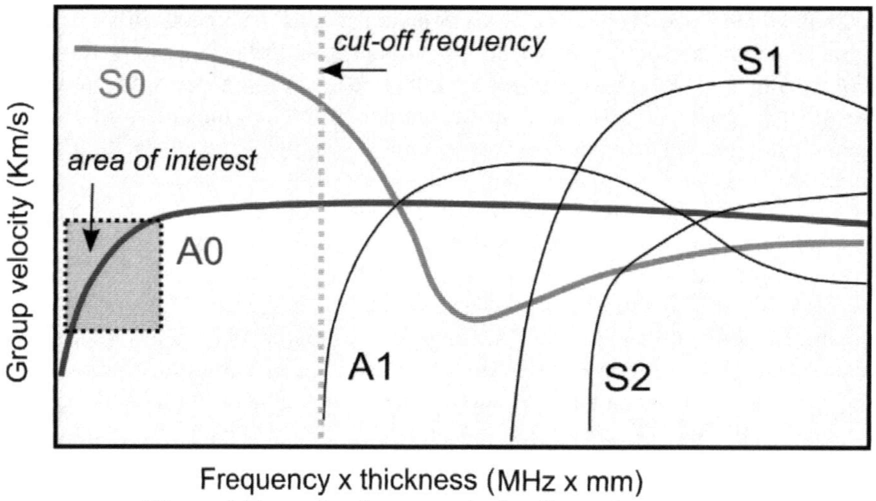

Figure 1 Example of group velocity dispersion curves.

As previously mentioned dispersion curves describe the relationship between the group velocity of different modes and the product of the actuation frequency and thickness of the structure. In the low values of the frequency-thickness product the velocity of A0 mode rapidly increases with the thickness change as is highlighted in the Figure 1. For the purpose of DCB measurements it is useful to choose the actuation frequency within this range. With delamination growth the part of the specimen with half thickness extends which results in time delays of signal arrivals compared to the baseline measurement.

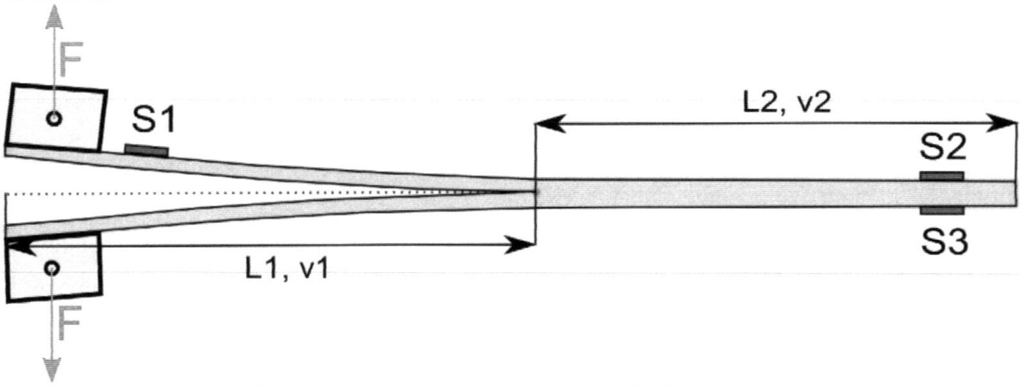

Figure 2 Sensor arrangement on the DCB specimen.

The basic arrangement is shown in Figure 2. Sensors S2 and S3 are mounted symmetrically on different sides of the specimens. If the sensor S1 is an actuator this arrangement enables to determine the specific propagating mode. Symmetrical mode reaches the sensors S2 and S3 with the maximum wave deflection at the same time so that the wave for the 2 sensors propagates in phase whereas asymmetrical mode would look out of phase. The asymmetrical mode A0 actuated by the sensor S1 and measured by the sensors S2 and S3 is shown in Figure 3 for the baseline measurement and for delamination of 74.5 mm.

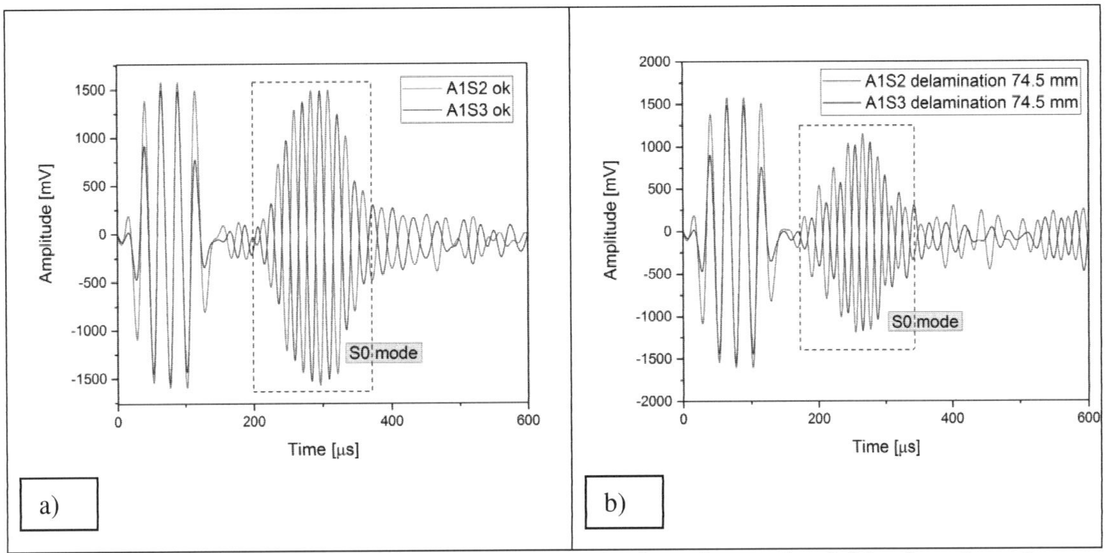

Figure 3 Actuated symmetrical mode A0 propagating in undamaged and delaminated specimen.

3. Results

An equation based on the known velocities for the both thicknesses and time of flight delays during the test was analytically derived as follows:

$$a_n = \frac{v_1 v_2 dt_n}{v_2 - v_1} \tag{1}$$

where v_1, v_2 are group velocities for the half thickness and the original thickness respectively, dt_n denotes difference in the time of flight compared to the baseline measurement and a_n represents delamination length.

Using the equation (1) delamination lengths were calculated by two different approaches. The first technique uses group velocities and time differences by tracking the maximum amplitude peak of the wave and the second uses the first arrival of the wave. Figure 4 gives an example of determination the time delays by means of these two approaches. Due to the dispersion character tracking the first arrival peak or the maximum amplitude peak show different time delays as shown in Figure 4 – 4.82 μs or 6.76 μs, respectively. In general tracking any peak would cause different delays.

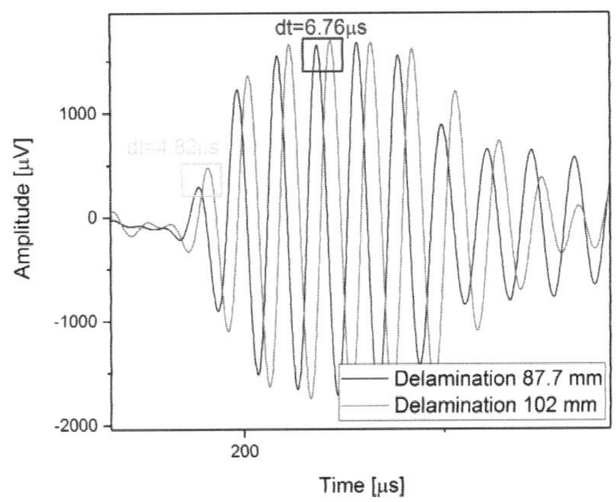

Figure 4 Example of time delays of the corresponding peaks for 2 delamination lenghs.

Dispersion and mode conversion on the crack tip may cause the change in the maximum amplitude peak and increased number of peaks compared to the baseline measurements. After the final failure the last measurement was conducted for the half thickness specimen. The same energy was actuated as into the specimen of the original thickness, so that the wavepacket contained more peaks. Therefore, tracking the corresponding peaks is essential. Example of all measurements is shown in Figure 5. With increase of the split part of the specimen the maximum amplitude peaks overflows to the earlier peaks and the content of other modes (converted or backwalled) is included into the wave packet.

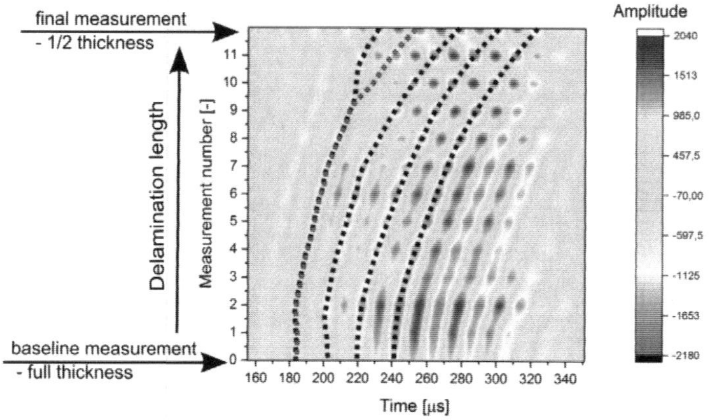

Figure 5 Time delays of different peaks of the wave for all measurements.

The comparison of delamination length using visually based tracking and Lamb wave measurements is shown in Figure 6 for 2 specimens. Calculating the mean absolute percentage error (MAPE) the approach of tracking the maximum peak amplitude shows the average error (average for all specimens) only 7.9 % while the approach using first peak arrival 13.7 %. As can be seen from Figure 6 b crack lengths determined visually might be underestimated.

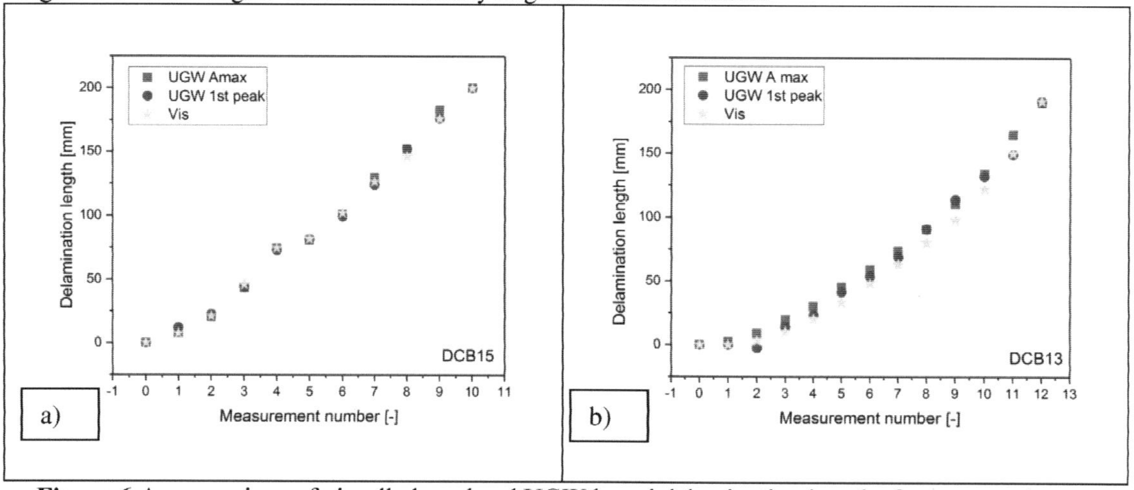

Figure 6 A comparison of visually based and UGW based delamination lengths for two specimens.

4. Discussion

The main concern of this paper was to exploit delayed Lamb wave time of flights during DCB test and calculate exact delamination lengths. Particular attention is paid to the suitable peak tracking. The following findings based on the experimental results can be stated.

- The suitable central frequency to actuate the A0 mode in the 3 mm thick specimen lies in the range of 40 - 60 kHz. Faster S0 mode is "hidden" in the crosstalk and A0 is therefore very well distinguished.

- The A0 mode propagates slower in the split part of the specimen and therefore time delays are recorded.
- If the crack tip approaches the sensor S2 (the split part of the specimen increases) probably another converted mode content amplifies and/or the same amount of actuated energy propagating in the longer part of the half thickness specimen (compared to the longer full thickness part) cause propagation of more peaks. In order to find out the exact reason numerical simulation would be useful to understand the propagation in this exact test.
- Based on the absolute mean percentage error of visually based and Lamb wave based measurements tracking the maximum amplitude peak showed better agreement. However, visual determination of delamination length might be underestimated. More accurate delamination lengths might lead to better agreement.

5. Conclusion

Based on the results it can be concluded that presented equation using time of flight and group velocity of Lamb wave A0 mode with the central frequency of 48 kHz can be used for determination of delamination length for DCB specimens. However, for the future study numerical methods would be beneficial and might help with the suitable frequency selection with respect to the peak tracking.

References

[1] Military handbook 1999 *USA, Handbook on Nondestructive evaluation system*
[2] Military Handbook 2012 *USA Composite Materials Handbook* Volume 3 Polymer Matrix Composites Materials Usage, Design, and Analysis
[3] Gholizadeh S 2016 A review of non-destructive testing methods of composite materials *Procedia Structural Integrity* 1 pp 50–57
[4] Kadlec M, Růžek R 2011 A Comparison of Laser Shearography and C-Scan for Assessing a Glass/Epoxy Laminate Impact Damage *Applied Composite Materials* 19(3-4) pp 393–407
[5] Rose J L 2014 *Ultrasonic guided waves in solid media* (New York: Cambridge Univ. Press)
[6] Michalcová L Růžek R 2016 Fatigue test of an integrally stiffened panel: Prediction and crack growth monitoring using acoustic emission. *Procedia Structural Integrity* 2 pp 3049–3056
[7] García I et al. 2015 Optical Fiber Sensors for Aircraft Structural Health Monitoring *Sensors* 15(7) pp 15494–15519
[8] Ruzek R et al. 2014 Strain and damage monitoring in CFRP fuselage panels using fiber Bragg grating sensors Part II: Mechanical testing and validation *Comp. Structures* 107 pp 737–744
[9] Cagáň J 2016 Hardware implementation of electrical resistance tomography for damage detection of carbon fibre–reinforced polymer composites *Structural Health Monitoring: An International Journal* 16(2) pp 129–141
[10] Giurgiutiu V 2016 *Structural health monitoring of aerospace composites* (London: Elsevier)
[11] Su Z Ye L & Lu Y 2006 Guided Lamb waves for identification of damage in composite structures: A review *Journal of Sound and Vibration* 295(3-5) pp 753–780
[12] Bonet M et al. 2015 Identification of Barely Visible Impact Damages on a Stiffened Composite Panel with a Probability-based Approach. *Structural Health Monitoring 2015*
[13] Salmanpour M S, Sharif Khodaei and Z Aliabadi M H 2017 Instantaneous Baseline Damage Localization Using Sensor Mapping *IEEE Sensors Journal* 17(2) pp 295–301
[14] Wang D et al. 2012 Monitoring of delamination onset and growth during Mode I and Mode II interlaminar fracture tests using guided waves *Composite Science and Tech.* 72pp 145–151
[15] Karpenko O Y et al. 2013 Lamb Wave Based Monitoring of Delamination Growth in Mode I and Mode II Fracture Tests. *Conf. Proceed. of the Society for Exper. Mech. Series* pp 33–43
[16] Ramadas C et al. 2012 Ultrasonic Lamb Wave Based Crack Growth Prediction for Estimation of Strain Energy Release Rate *Advanced Materials Research* 585 pp 24–28

Simulation and Analysis of One-time Forming Process of Automobile Steering Ball Head

Shi Peicheng*, **Zhang Xujun, Xu Zengwei and Zhang Rongyun**

Anhui Polytechnic University, Anhui Engineering Technology Research Center of Automotive New Technique, Wuhu 241000,China

E-mail:shipeicheng@126.com

Abstract. Aiming at the problems such as large machining allowance, low production efficiency and material waste during die forging of ball pin, the cold extrusion process of ball head was studied and the analog simulation of the forming process was carried out by using the finite element analysis software DEFORM-3D. Through the analysis of the equivalent stress strain, velocity vector field and load-displacement curve, the flow regularity of the metal during the cold extrusion process of ball pin was clarified, and possible defects during the molding were predicted. The results showed that this process could solve the forming problem of ball pin and provide theoretical basis for actual production of enterprises.

1. Introduction

Ball pin is an important part of the automobile steering system, its strength and reliability directly concerns steering sensitivity, driving safety of the automobile. Therefore, a wear-resisting ball surface and a heart with certain strength and toughness are required to withstand impact load. Ball head parts processing & manufacturing by ordinary machinery features great difficult, low efficiency and serious material waste. Cold extrusion technology is a plastic forming technology with high production efficiency and material utilization rate. In the process of extrusion, the metal deformation part is subjected to three-way compressive stress, so grain structure of the material is more compact and surface quality of the workpiece is fine[1]. Therefore, forming of ball pin parts by cold extrusion can not only save metal materials, achieve near-net forming, but also improve the ball pin size, surface accuracy and mechanical strength.

In this paper, analog simulation study of cold extrusion forming process of ball pin was made by finite element numerical simulation, the flow regularity of the metal during the cold extrusion process of ball pin was analyzed, and possible defects during the molding were predicted, to reduce industrial production cost of ball pin parts by cold extrusion, improve the product quality and provide theoretical basis for the actual production of enterprises[2-4].

2. Analysis of ball head forming process and die design

For large size ball head pin, the ball head and the ball arm sizes are greatly different. As shown in **Figure 1**, the ball head has a diameter of 38mm, the arm has a minimum diameter of 22mm. Made of

IOP Conf. Series: Materials Science and Engineering **326** (2018) 012007 doi:10.1088/1757-899X/326/1/012007

material 40Cr, the ball pin belongs to long shaft type part. According to **Figure 1**, such part has an inverted cone structure in radial direction, so extrusion is difficult. To solve this problem, this paper adopted combined die structure molding. As shown in **Figure 2**, two pairs of combined dies are symmetrically arranged, with the bonding accuracy kept by the guide plate fixed on the jacket. The two pairs of female dies are machined with corresponding grooves so that sliding along the guide plate is possible. The two pairs of female dies are closed first before the extrusion forming to form a closed space, and then male die continues to go downward for extrusion forming. After the extrusion forming is completed, the push rod pushes the extrusion part and the combined die upward together, so that the combined die slides along the guide plate to achieve lateral die splitting. Then, remove the part to complete one working cycle[5-7]. As the ball pin parts are all rotary parts with axial symmetry, 1/4 ball pin is analyzed in this paper to render analytical calculation amount small.

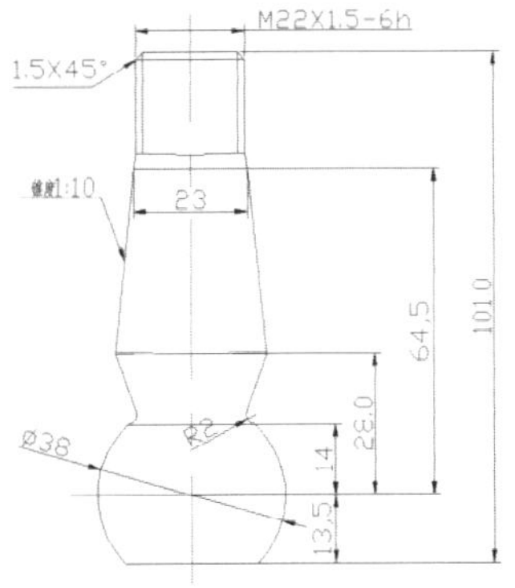

Figure 1. ball pin parts

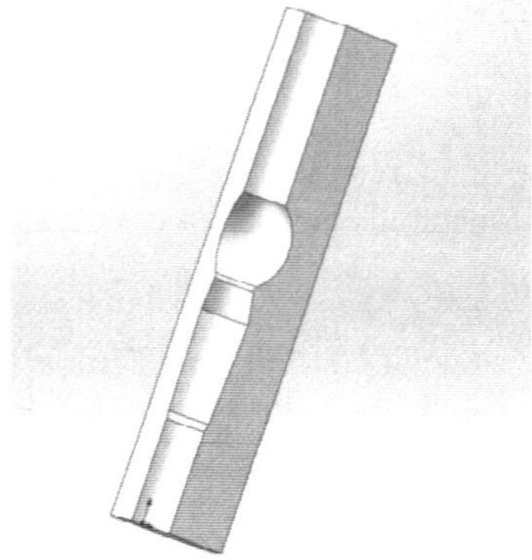

Figure 2. female die

3. Blank dimension calculation

The blank volume is calculated according to the size in the drawing of extrusion part in the principle that the blank volume is equal to the volume of the extrusion part. If machining is needed after cold extrusion, trimming allowance should be added when calculating the blank volume. Main considerations should be given to deformation degree, height-diameter ratio and diameter and length of the thread part during calculation. The larger the blank diameter is, the more stable the extrusion deformation in the die cavity is, and the less difficult it is in ball head forming, but formation of the inverted cone part will be affected. According to experience and calculation, the diameter of 22 mm at the journal is chosen as the reference size of the outer diameter of the blank. The blank length is calculated according to the principle of volume invariance. Finally, the calculated blank length is 165mm.

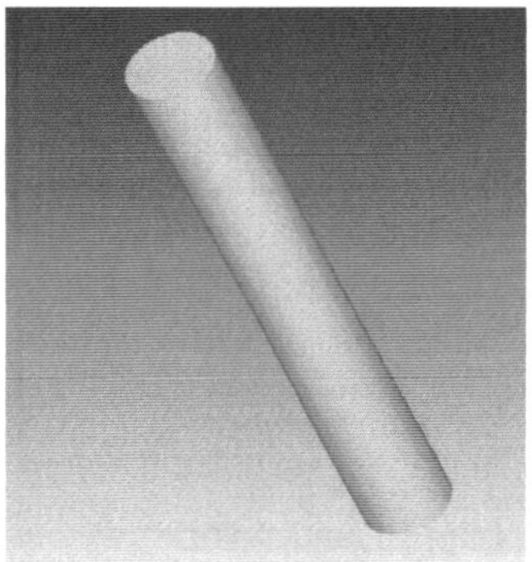

Figure 3. blank

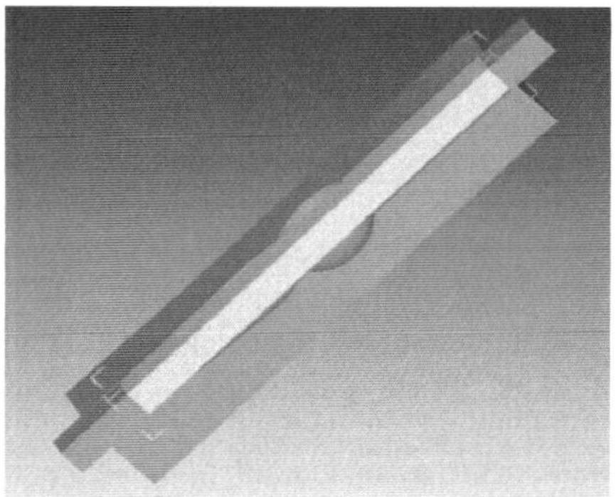

Figure 4. extrusion forming chart

4. Blank pre-treatment

According to the cold extrusion process analysis, softening process, surface and lubrication treatment is a must for blank materials before cold extrusion forming.

4.1. Softening of blank material

Alloy structural steel 40Cr features high hardness, big deformation resistance and poor plasticity in the supply status. It will be difficult if direct cold extrusion forming is performed without softening annealing, not only making forming parts fail to meet the requirements, but also possibly damaging the die. Therefore, spheroidizing annealing is a must before cold extrusion process to improve the plasticity, reduce the hardness and deformation force[8-9]. That is, first heat up to 800°C, keep for 12 hours, and then cool with the furnace to below 500°Cat a speed of 150°C/ h, followed by air cooling to obtain granule microstructures for improved plasticity and reduced hardness. After the softening treatment, 40Cr alloy steel blank hardness is 179HB, tensile strength can be reduced from above 980MPa to below 590MPa to facilitate metal flow forming in the extrusion forming.

4.2. Surface treatment and lubrication

Surface treatment is a key process in the cold extrusion process, which has a significant effect on the surface quality and die life of the extrusion part. Cold extruding steel surface is subjected to phosphating saponification treatment to reduce the metal flow and die friction in the molding process, cut down the extrusion force and improve die life.

5. Simulation of Cold Extrusion Forming Process

5.1 Establishment of finite element model

Build the model graph of ball head forming female die(**Figure 2**) and ball pin blank(**Figure 3**) in the three-dimensional software Solidworks, to be respectively saved and then separately saved as .stl format file.

Open the finite element analysis software Deform-3D, import the blank (.stl format file) for analysis. Set the blank parameters, with elastic modulus of hardening and tempering 40Cr at 2.05 × 105MPa, Poisson's ratio at 0.29, and density at 7.85 g/cm^3. The grid is divided into 24887 smaller grids as shown in **Fig.5**

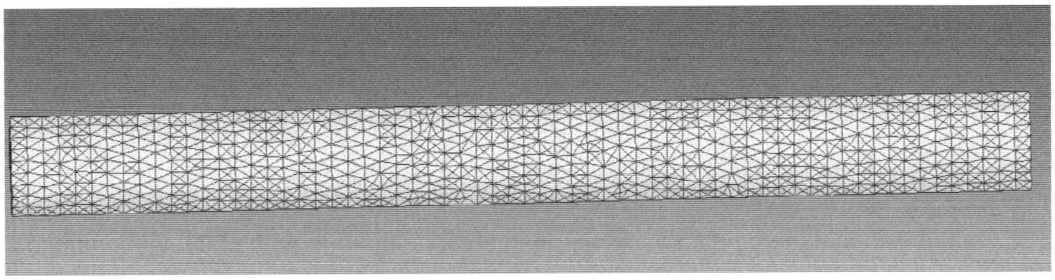

Figure 5. grid partition diagram

Import the die (See **Figure 4**), check the geometry of the die, set its parameters and adjust its position. Then, add the contact relation between movement blank and die, with friction coefficient set at 0.2 and temperature set at 20℃. Then, set the die movement direction as -y, set the movement speed of top die at 1 mm/sec, the distance between the top and lower die at 173mm, set the total number of steps at 145 steps to be saved once every 5 steps, and set the die increment at 0.5mm/sec. Check and generate DB files, run DB files, and wait for computer analog simulation results[10].

5.2 Simulation of ball pin extrusion process and result analysis

During the analog simulation, velocity field, equivalent stress strain, load -stroke relationship is analyzed as follows[11-14]:

5.2.1Analysis of velocity vector field. As shown in **Fig.6**, the left and right female dies are closed first, and then male die goes downward after the start of ball pin extrusion forming. Under the compression of male die, the blank moves downward, generating downward initial velocity field. In the beginning, the metal bottom has the fastest flowing speed. As the blank diameter is equal to the journal width, position with fast speed from the start to Step 15 first contacts with the female die. At this moment, there is still a short distance between the blank and the bottom of the female die. The upper metal makes rigid downward shift, while the bottom metal forms chamfering shape. At Step70, the blank thoroughly contacts with the bottom of the female die, with the blank entering upsetting stage. Metal at the journal has significantly faster flowing speed at this time, which is because the metal flow is always in the direction with the minimum resistance. Ultimately, the female die is filled at Step130. Then, the male die continues downward movement, the metal flows uniformly in the enclosed space, forming inverted cone shape in radial direction (as in Step145).

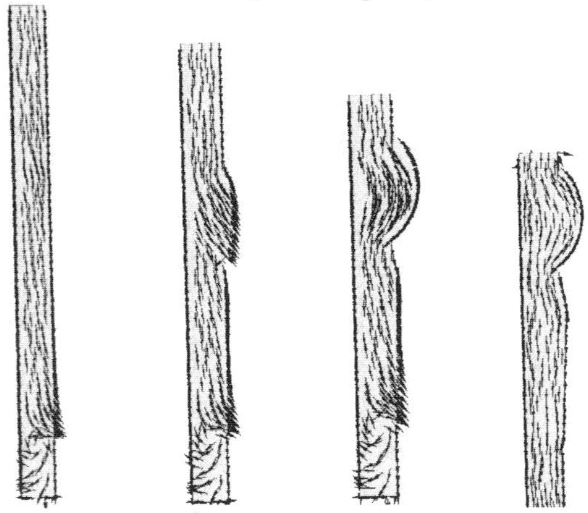

Step15 Step70 Step130 Step145

Figure 6. diagram of velocity vector field

At different stages of metal forming, the velocity field shows that the metal deformation process is smooth, the surface has no folding, breakage defects, and metal flow stops after the completion of forming, which is conducive to the part stability and die life.

5.2.2Equivalent stress and strain analysis. The stress diagram of the cold extrusion forming is shown in **Fig.7**.It can be seen from the fig.7 that the maximum equivalent stress of the extrusion forming occurs where the outer edge of the ball head and inverted cone is about to be filled. As the male die moves downward, the blank fills the entire female die, with the stress shifted to the male die inlet, journal rounded corner and the bottom chamfer. The outer edge and inverted cone of the ball head is subjected to great deformation, large stress, thus prone to wear, cracking. According to the forming deformation, the stress and strain trajectories of extrusion are studied by 9 points of P1~P9. As can be seen from the graph in **Fig.7**, point P3 is the maximum stress point, with value at 910Mpa; followed by points P4, P7, P6, which indicates that outer edge and inverted cone of the ball head withstands larger stress during forming; Point P4 rises sharply in the end, indicating that the final forming part is the outer edge of the ball head.

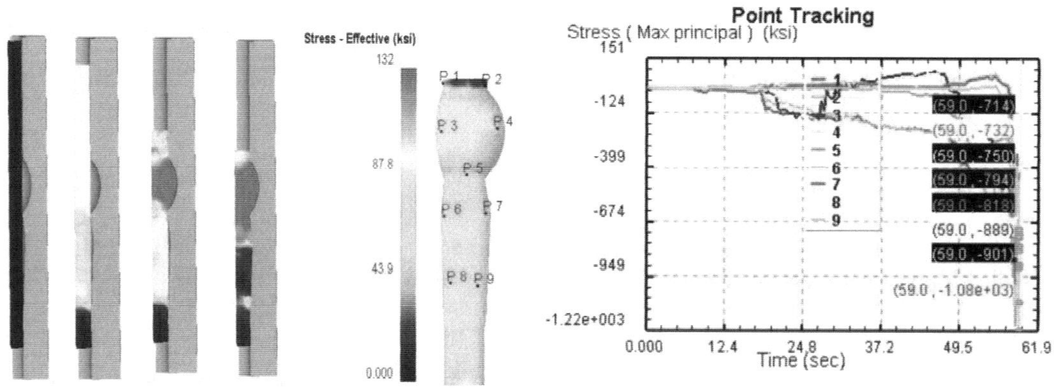

Figure 7. Stress diagram of extrusion forming

In the strain diagram of extrusion forming as shown in **Fig.8**, when the extrusion starts, the bottom chamfer tends to be zero, and the strain is layered distributed from the outer edge to the center of the ball head during upset-extruding. The maximum strain appears at the finally extruded inverted cone, which is because the flow direction of metal above the journal is against the inverted cone should be formed, then the maximum equivalent strain with value at 2.01 generates in between. The maximum strain at the time of final forming occurs at the journal turning, which is because metal of the formed ball head needs further downward flow to form the cone structure, and then the metal at the heart of the journal withstands excessive tensile stress and additional stress. According to the forming deformation, the stress and strain trajectories of extrusion feature points are studied by 9 points of P1~P9. It can be seen from the graph that, point P3 withstands the largest strain, which is because the core is subjected to axial compressive stress and radial additional tensile stress; the next is point P5 which is the last point with largest strain. This is because the formed ball head forces the metal to further flow downward to form a cone structure; point P4 rises sharply in the final forming, indicating that the ball head is the final part to complete forming. Points P6, P7, P8, P9 withstand relatively gentle strain with small deformation, indicating normal flow of metal.

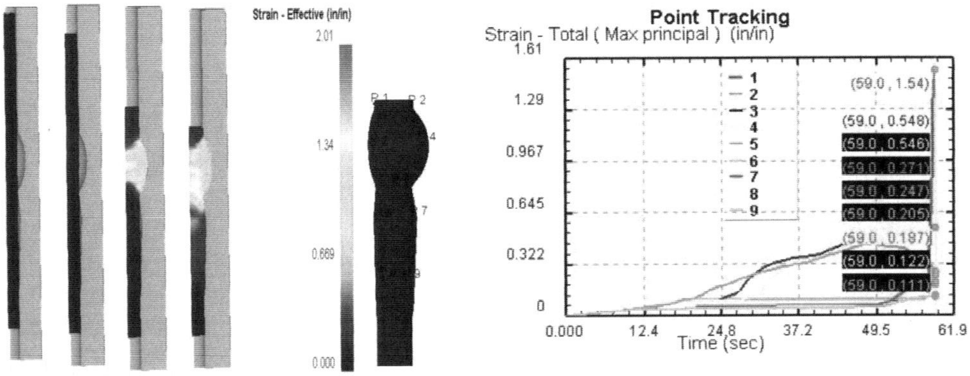

Figure 8. Strain diagram of extrusion forming

As shown in **Figure 9**, the load-displacement diagram demonstrates that the load is stable throughout the forming process, there is no obvious fluctuation and the load is also small, which is very beneficial to improvement of the die life. The load rises sharply after forming is basically completed in later stage of forming, which, however, does not exceed the bearing load of the die.

As shown in **Fig.10**, the wear coefficient diagram indicates that the wear coefficient is increasing as a whole, and as the extrusion process progresses, workpiece wear increases due to uneven deformation of the workpiece, residual stress and growing metal lattice distortion. It can be seen from the **fig.10** that point P3 has the largest wear coefficient, followed by points P4, P5, indicating that these parts are prone to destruction for ball pin.

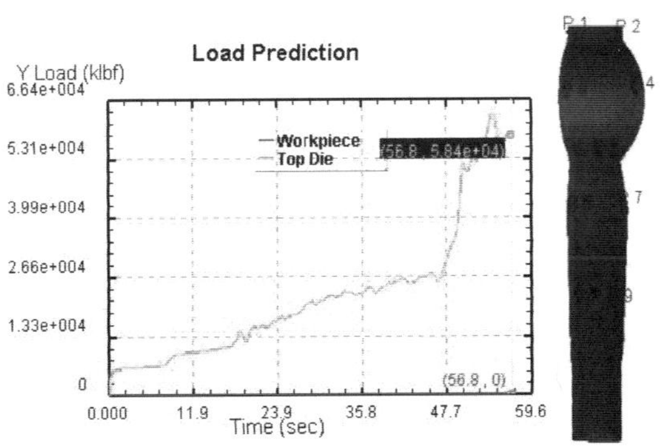

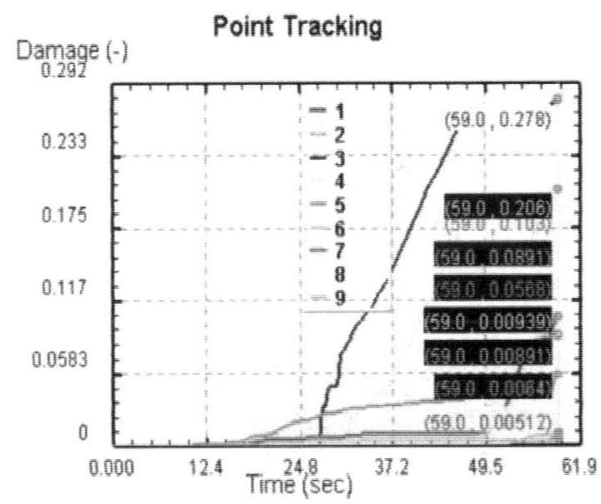

Figure 9. Load － Displacement diagram **Figure 10.** Wear coefficient diagram

Comprehensive analysis of the above concludes that the ball head axis withstands the maximum stress in the forming process, while the inverted cone withstands the maximum strain. The ball pin maintains stability in the extrusion process. The extrusion force has been in steady growth, and the load rises quickly before the process completion, which is in line with the law of rheology; however, the extrusion force remains within a reasonable range, not beyond the bearing load of die.

6.Conclusion

(1) The use of closed upset-extruding, one-time extrusion forming of combined die can better solve problems of large-scale ball pin in big machining allowance, low utilization of materials and low production efficiency.

(2) Through analysis of structure of the ball pin, a reasonable production process was selected. The analog simulation of the forming process was carried out by using the finite element analysis software Deform-3D to find out the flow regularity of the metal during the cold extrusion process of ball pin.
(3) The simulation can effectively shorten the product development cycle and reduce the cost. Providing some theoretical basis for the actual industrial production of ball pin, it has certain research and guiding significance.

Acknowledgments
The authors would like to thank the financial supports of the National Natural Science Foundation of China (Grant No. 51575001,51605003), of Anhui province science and technology research key project (Grant No. 1604a0902158), of Anhui university scientific research platform innovation team building projects (2016-2018), and of Anhui Polytechnic University Young and middle-aged top-notch talent project (2016BJRC010).

References
[1] M. Plancak,Branimir Barisic,Branko Grizelj. Different Possibilities of Process Analysis in Cold Extrusion[J]. Key Engineering Materials, 2008, 367:209-214.
[2] Shao Rong Yang,Yun Feng Lu. Study on Cold Extrusion Technology of 20Cr Bevel Gears[J]. Advanced Materials Research, 2012, 548:586-590.
[3] Shi Min Xu,Hua Gui Huang,Deng Yue Sun.Precision Forming of Pipe Joint Based on Cold Extrusion Technology[J]. Applied Mechanics & Materials, 2010, 37-38:875-879.
[4] Zhi Hong Yu. The Research of the Cold Extrusion Shaping Technology of the Tubular Component with Oral Part Having Flange[J]. 2012, 4:121-124.
[5] Hong Wang.The stamping and mold design on stainless steel cover[C]// Second International Conference on Mechanic Automation and Control Engineering. IEEE, 2011:5433-5436.
[6] Cao Suhong.3D mold design of outlet based on pro/ engineering[C]// Symposium on Electrical & Electronics Engineering (EEESYM).IEEE,2012:39-43.
[7] Zhang Qingping; Cui Huanyong; Wang Yuzeng.Process Design for Cold Precision Forging of Bevel Gear[C]//International Conference on Digital Manufacturing & Automation.IEEE, 2010:114 -116.
[8] Geun An Lee,Seo Gou Choi,Dong Jin Yoon,Hee Woong Lee,Kyoung Hoan Na. Forming Technology for Cold Forging Processes of Ball Stud Using Non-Heat-Treated Cold Forging Materials[J]. Materials Science Forum, 2005, 475-479:3247-3250.
[9] B. Suthep; K. Uten.Cold forging deformation analysis for stainless materialusing DEFORM[C] //International Conference on Industrial Engineering and Engineering Management (IEEM). IEEE , 2015:1277 - 1281.
[10] Yu Qin Li,An Min Zhang.Wear Analysis of Warm Extrusion Dies Using DEFORM[J]. Applied Mechanics & Materials, 2014, 529:630-635.
[11] M.N.Burcham,R.Escobar,C.O.Yenusah,T.W.Stone,G.N.Berry,A.L.Schemmel,B.M.Watson,C.U. Verzwyvelt.Characterization and Failure Analysis of an Automotive Ball Joint[J].Journal of Failure Analysis & Prevention, 2017:1-13.
[12] Arbtip Dheeravongkit,Narongsak Tirasuntarakul. Parametric Study of Swage Ball for Hard Disk Drive Swaging Process Using Finite Element Method[J]. Applied Mechanics & Materials, 2013, 277:2241-2247.
[13] Tomasz Bulzak,Zbigniew Pater,Janusz Tomczak. Numerical and experimental analysis of a cross wedge rolling process for producing ball studs[J]. Archives of Civil & Mechanical Engineering, 2017, 17(4):729-737.
[14] Srikar Potnuru,Raviteja Vinjamuri,Susant Kumar Sahoo,Santosh Kumar Sahoo. Three Dimensional Analysis of Combined Forward and Backward Extrusion-Forging Process Using DEFORM 3D[J]. Applied Mechanics & Materials, 2014, 592-594:791-795.

IOP Conf. Series: Materials Science and Engineering **326** (2018) 012008 doi:10.1088/1757-899X/326/1/012008

Study on the Weak Stress in Flexural MEMS Cantilever

Yuetao Ge[1] and Yan Ren[2]

Beijing HIWING Scientific and Technological Information Institute, Beijing, 100074, China.

[1]geyuetao@gmail.com, [2]renyan1979@outlook.com

Abstract. In order to design a better piezoresistive MEMS cantilever beam, especially for cantilever beams that will detect weak forces or will be subjected to weak forces, this paper uses study on the weak stress in flexural MEMS cantilever. The sensor design structure, divided into protective layer, piezoresistive layer, support layer. The protective layer is responsible for protecting the piezoresistive layer so that the varistor is insulated from the outside; the piezoresistive layer is used to make the varistor; the support layer forms the main part of the cantilever beam, the majority of the cantilever beam. This paper has some value for cantilever multilayer structure design and cantilever beam size design.

1. Introduction

In recent years, biomolecules and their unique ability of recognition have been investigated in terms of their mechanical response such as quartz crystal microbalance (QCM) to external forces. Biological or chemical MEMS cantilever biosensors and biochips develop after atomic force microscope (AFM), which is a common feature of methods of mechanical response by applying external force, invention and employ using MEMS technology [1]. The AFM consists of a cantilever with a sharp tip (probe) at its end that is used to scan the specimen surface detected. The cantilever is typically silicon or silicon nitride with a tip radius of curvature on the order of nanometers. When the tip is brought into proximity of a sample surface, forces between the tip and the sample lead to a deflection of the cantilever according to Hooke's law. On this basis, forces that are measured in AFM include mechanical contact forces, Van der Waals forces, capillary forces, chemical bonding, electrostatic forces, magnetic forces, Casimir forces, solvation forces, etc. [2]. And the principle of BioMEMS cantilever biosensor or biochip is that MEMS cantilever surface combines probe molecules after biochemical treatment [3]. While target molecules in the test sample emerge hybridization reaction or other biochemical reactions with probe molecules, the stress on the cantilever surface changes and the cantilever bending deformation occurs, and then it can obtain reaction information through testing deformation of the cantilever [4]. Lang laboratory first reported that they take cantilever into DNA detection in *Science* 2000 [5]. They converted the weak signals of biochemical reaction to directly and specifically transducer molecular recognition into nanomechanical responses in a cantilever array. MEMS cantilever piezoelectric biosensor is reported most, but the signal is interfered by noise easily and hard to judge whether resonance while instability and detect piezoelectric signal while stability.

In this paper, we study a MEMS cantilever under a concentrated, transverse loading force, which applied at the free end. We often meet this situation in MEMS design. The transverse force loading comprises both longitudinal strains and shear strains. We ignore the shear stress in this paper.

2. Analysis of the stress in flexural MEMS cantilever

Under a transverse loading of a concentrated force at the free end, the torque distribution through the length of the beam is non-uniform—it is zero at the free end and reaches a maxim at the fixed end. At any cross section, the signs of the longitudinal stresses change across the neural axis. The magnitude of stresses at any point on the cross section is linearly proportional with respect to the distance to the neutral axis. The distribution of stress in MEMS cantilever is described in Figure 1.

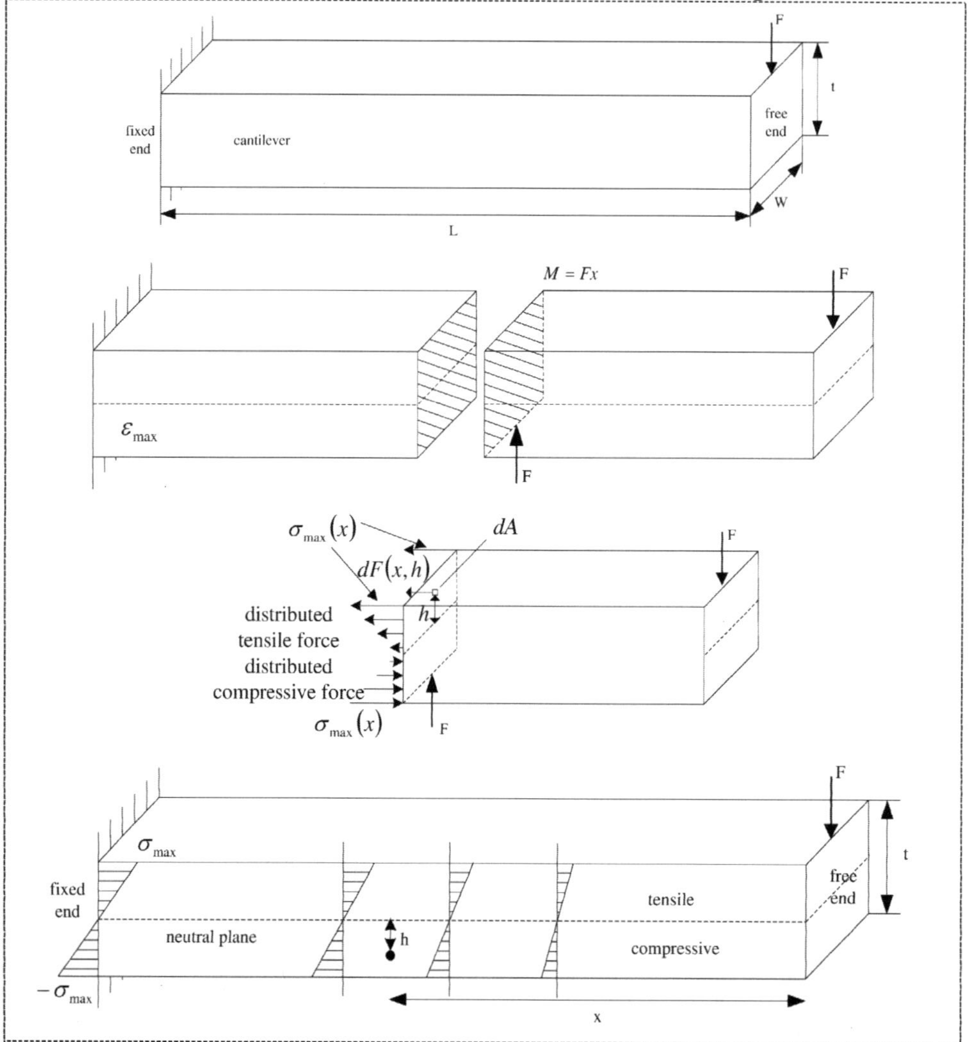

Figure 1. Stress distribution in a uniform and symmetric MEMS cantilever beam

The magnitude of the maximum stress is related to the linear variation of the cross-section with respect to the free end of the distance, reaching the maximum width at the top and bottom ends. In general, the varistor is usually selected in the cantilever beam surface and near the fixed end.

The length of the cantilever beam (L) is that the axis x begins at the free end to the fixed end. The positive stress at any given cross-section (located at axis x) and the distance of the neutral plane (h) is expressed as $dF(x,h)$. The total reaction torque of the cross-section is that the area integral dA (ie $dF(x,h)$) of orthogonal force (normal force) acting on any given area multiplied by the distance between the force and the center of the arm, ie,

$$M = \iint_A dF(x,h)h = \int_w \int_{h=-\frac{t}{2}}^{\frac{t}{2}} (\sigma(x,h)dA)h \tag{1}$$

Assuming that the magnitude of the stress is linearly related to h, and it has a maximum value ($\sigma_{max}(x)$) on any cross-sectional surface, the moment balance equation at any given cross-section is shown as:

$$M = \int_w \int_{h=-\frac{t}{2}}^{\frac{t}{2}} \left(\sigma_{max}(x) \frac{h}{\left(\frac{t}{2}\right)} dA \right) h \tag{2}$$

The maximum pull of the cantilever occurs at the fixed end (x = L). Generally, the only interest is the maximum pull value and the pressure value of the fixed end in the design. According to the total torque, the maximum tension expression is

$$S_{max} = \frac{M(x)t}{2EI} = \frac{FLt}{2EI} \tag{3}$$

3. Multilayer Piezoresistive MEMS cantilever Structural Design

The varistor has a fixed length and width, and if the resistance is formed by doping the silicon beam, the piezoresistive element will be located below the upper surface (Figure 2-a). If the resistance is the formation of deposited polysilicon or metal layers, the surface of the piezoresistive element is above the upper surface (Figure 2-b). In both cases, the resistors begin at the root of the cantilever beam. Assuming that the piezoresistive resistor is thin in the vicinity of the upper surface of the cantilever and is shorter than the length of the cantilever beam, the resistance of the resistor can be approximated by the formula (3).

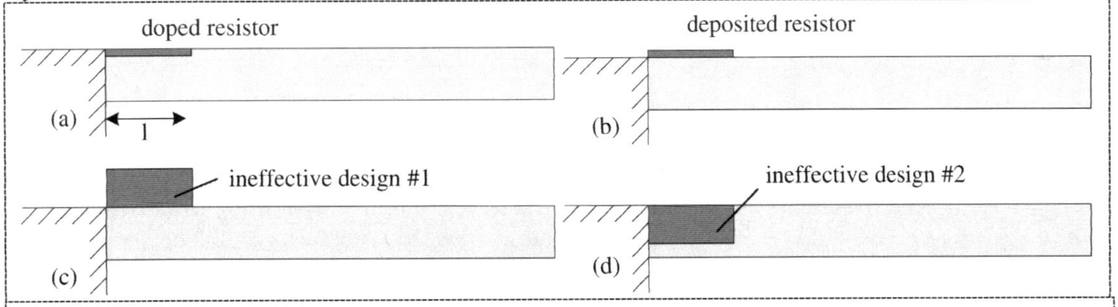

Figure 2. Design position of cantilever varistor

The varistor is too thick to be designed and manufactured desirably, as shown in Figure 2-c and Figure 2-d. In Figure 2-d, the doped varistor is very deep below surface in the cantilever beam, if the doped area reaches below the axis, the sensitivity of the varistor is suppressed by the varistor below axis actually. In the extreme case, if the doped varistor covers the thickness of the entire cantilever, the resistance offsets each other due to compression and stretching of varistor. In Figure 2-c, if the doped varistor is too thick on the surface of the cantilever, the varistor needs to be considered for the cantilever as an integral part, the central axis position will change. The formula (3) needs to reconsider the varistor and the material of the beam. From an electronic point of view, the thick resistors have no

IOP Conf. Series: Materials Science and Engineering **326** (2018) 012008 doi:10.1088/1757-899X/326/1/012008

any advantage over fabrication because they exhibit relatively weak impedances and also require a longer processing time (doping or deposition).

In this paper, the composite multi-layer cantilever beam sensor is applied for structure design, divided into protective layer, piezoresistive layer and support layer. The protective layer is responsible for protecting the piezoresistive layer so that the varistor sensor is insulated from the outside; the piezoresistive layer is used to make the varistor; the support layer forms the main part of the cantilever beam, i.e., the majority of the cantilever beam.

The four completely identical cantilever beams are taken as a group, with reference to the characteristics of the universal glass biochip, each homologous gene needs to be tested for repeated gene array measurements. Therefore, at the same time, the four cantilever beams are assigned to each group, and each group is used to measure the same gene. One of the cantilever beams in each group is used as the beam to be tested and the other is the same cantilever beam as the reference beam. A symmetrical Wheatstone bridge circuit is formed with four identical varistors, two of which are located at the same position on the tested beam and the reference beam respectively, and the other two on the substrate. The presence of factors such as external ambient noise, interference, thermo-mechanical vibration noise, or temperature changes may cause deformation of the cantilever beam, which can eliminate the signal error caused by the reference beam.

In the analysis of this paper, the effect of shear, residual stress, manufacturing process on cantilever beam is neglected, and the electrostrictive force is also neglected. Assuming that each layer is fully elastic, the thickness of the cantilever showed in Figure 3 is much smaller than the bent z-direction displacement of the cantilever caused by the pressure, so only the force of the vertical x and y-direction plane can be considered.

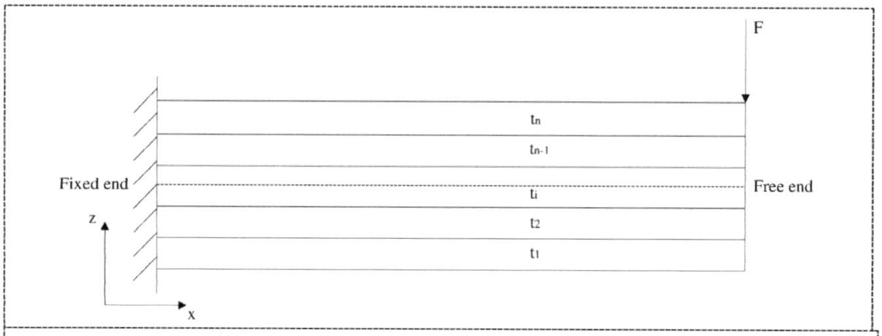

Figure 3. Multi-layer cantilever beam diagram

4. Simulation of weak stress in flexural MEMS cantilever

The simulated cantilever beam uses silicon as the material, the cantilever beam size is 400μm × 100μm × 10μm, the left end is fixed, and the right end is free. The parameters of silicon are: Young's modulus E = 170 × 109Pa, Poisson's ratio ν = 0.28, density ρ = 2329kg / m3. The gravity of the cantilever itself results in cantilever beam stress, the application of COMSOL Multiphysics (Femlab) simulation obtains:

The simulation of the cantilever beam stress caused by the gravity of the cantilever beam is shown as below in Figure 4.

The simulation of the displacement of the cantilever beam due to the gravity of the cantilever beam is shown as below in Figure 5, and the software is shown as exaggerated deformation.

IOP Conf. Series: Materials Science and Engineering **326** (2018) 012008 doi:10.1088/1757-899X/326/1/012008

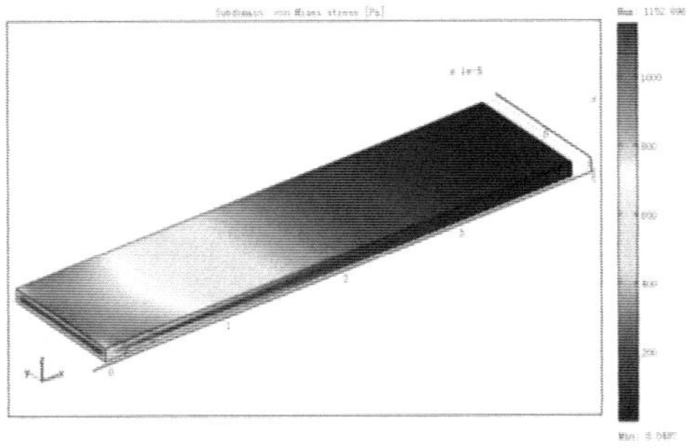

Figure 4. Stress simulation of cantilever beam caused by gravity of cantilever beam

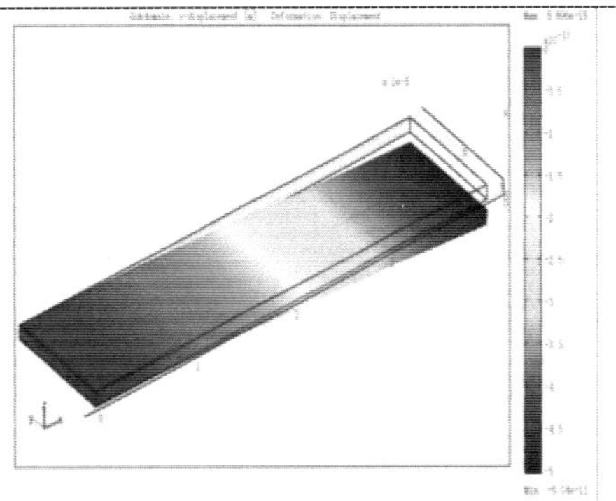

Figure 5. Simulation of cantilever displacement caused by self gravity of cantilever beam

The stress equipotential surface of the cantilever beam caused by the gravity of the cantilever beam to be simulated is shown in Figure 6.

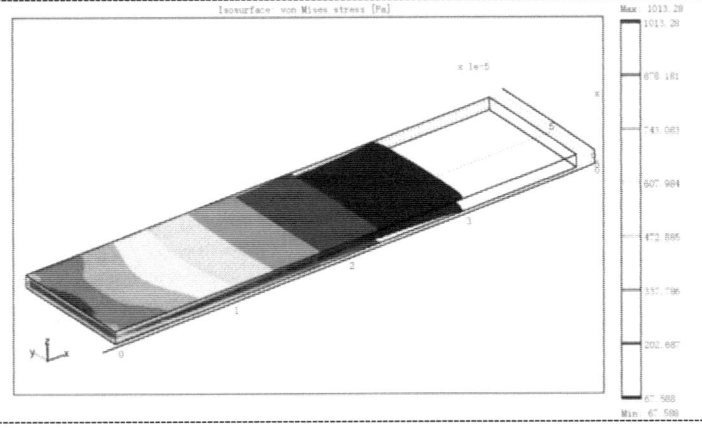

Figure 6. Equipotential surface of cantilever beam stress

Its maximum value is on the outside of the fixed end, and the free end of the above figure is enlarged as shown in Figure 7.

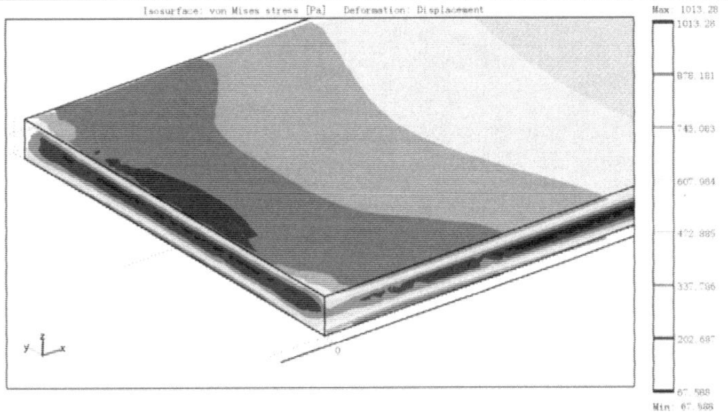

Figure 7. Partial amplification of cantilever beam stress equipotential surface of fixed end

The partial amplification of domain of fixed end is shown in Figure 8.

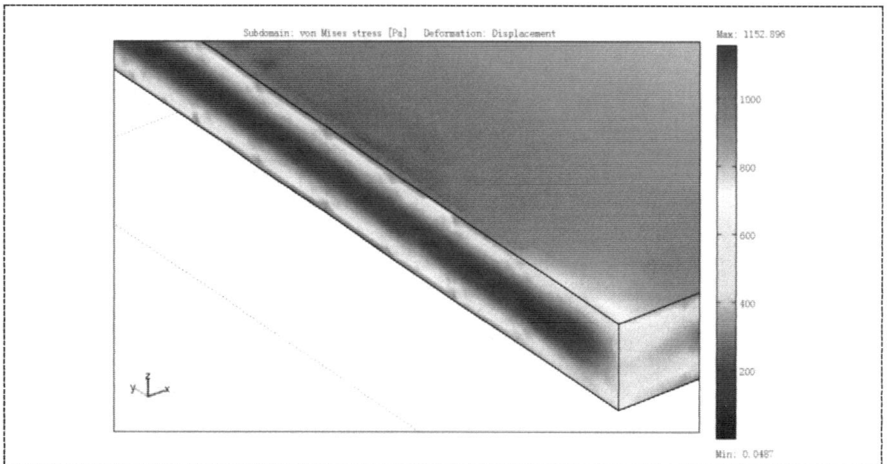

Figure 8. Partial amplification of cantilever beam stress domain of fixed end

It can be seen that the stress in the cross section of the cantilever beam is also stratified, and it will be greater when the stress goes outside.

5. Conclusion

This article studied the weak stress in flexural MEMS cantilever, which can be used on biosensors or biochips, and even on inertial navigation devices. In the case of MEMS cantilever beam (which will impose a weak force) design, according to the analysis of this paper, MEMS cantilever beam can be designed as the optimal multi-layer structure, and can be designed as the optimal size.

Reference

[1] Maria Gaczynska, and Pawel A. Osmulski, AFM of biological complexes: What can we learn? *Current Opinion in Colloid & Interface Science*, vol. 13, no. 5, 2008. pp. 351-367.

[2] Zhongliang Deng, Yuetao Ge, Weiguo Guan, Naibo Zhang, Qike Cao. Nanogold aggravating MEMS cantilever capacitance detection biochip [J], *Advanced Materials Research*. 2011. (159). 429-433.

[3] Algre E., Legrand B., Faucher M., Walter B. and Buchaillot L., Surface microscopy with laserless MEMS based AFM probes, *Micro Electro Mechanical Systems (MEMS), 2010 IEEE 23rd International Conference on.* Wanchai Hong Kong, 2010. pp. 292-295.

[4] Zhongliang Deng, Yuetao Ge, Qike Cao, Ke Han. The detection of a transgenic soybean biochip using gold label silver stain technology [J], *Bioorganic & Medicinal Chemistry Letters.* 2011. 21(22). 6905-6908.

[5] J. Fritz, M. K. Baller, H. p. Lang, H. Rothuizen, P. Vettiger, E. Meyer, H. –J. Güntherodt, Ch. Gerber, J. K. Gimzewski, Translating Biomolecular Recognition into Nanomechanics, *Science* 288(5464), 2000. pp. 316-318.

IOP Conf. Series: Materials Science and Engineering **326** (2018) 012009 doi:10.1088/1757-899X/326/1/012009

Modal Characteristics of Novel Wind Turbine Rotors with Hinged Structures

Hongya Lu[1, *], Pan Zeng[2] and Liping Lei[2]

[1]Beijing Institute of Mechanical Equipment, No.50 Yongding Road, Haidian District, Beijing 100039, China
[2]Key Laboratory for Advanced Material Processing Technology of MOE, Department of Mechanical Engineering, Tsinghua University, Beijing 100084, China

*subred@126.com

Abstract. The vibration problems of the wind turbine rotors have drawn public attention as the size of wind turbine has increased incredibly. Although various factors may cause the vibration problems, the flexibility is a big threat among them. Therefore, ensuring the high stiffness of the rotors by adopting novel techniques becomes a necessity. The study was a further investigation of several novel designs regarding the dynamic behaviour and the influencing mechanism. The modal testing experiments were conducted on a traditional blade and an isolated blade with the hinged rods mounted close to the root. The results showed that the rod increased both the modal frequency and the damping of the blade. More studies were done on the rods' impact on the wind turbine rotor with a numerical model, where dimensionless parameters were defined to describe the configuration of the interveined and the bisymmetrical rods. Their influences on the modal frequencies of the rotor were analyzed and discussed.

1. Introduction

The flexibility of large wind turbine rotors may lead to the vibration instability, such as the flutter problem, which is caused by the coupled vibration of flap and torsion [1, 2]. While the lightweight technique and advanced manufacturing process brought in large amounts of benefits, they resulted in a decreased inherent torsion frequency of the wind turbine rotor [2]. The decreased frequency is usually an indirect factor for the coupled vibration in the different directions. During the design of the rotors, the stiffness should be guaranteed to avoid the occurrence of the flutter vibration. In addition, the edgewise vibration instability is an urgent problem when the size of the rotors grows significantly, since the edgewise damping is usually smaller than that in the flap direction [3]. Therefore, the structural design of the wind turbine rotor should give consideration to the load bearing capability, weight, stiffness, damping, etc. at the same time. The stiffness of the rotor is the focus in this paper.

In the traditional design, the blade comprises an aerodynamic shell and the main spar. This spar usually has a box or "D" shape made of composite plies and a sandwich material to ensure bending stiffness. [4] Other designs adopted a new airfoil to ensure the aerodynamic behavior and the stiffness requirement, such as the flat-back foil by cutting part of the trailing edge, blunt trailing edge by continuously increasing the thickness of shell. [5] In addition, Roth-Johnson et al. proposed a biplane root-type blade and numerically studied the mechanical behavior of this structure.[6, 7] They showed that a biplane structure has a smaller tip displacement, higher lift–drag ratio. [6, 7]

The modal frequency reflects the dynamic stiffness of the structure. Frequency maximization [8] incorporates the optimized constraints of the segmented spar length to satisfy both the weight and stiffness requirements. [9] Tarfaoui et al. used the finite element method to study the influence of the spar configuration on the modal frequency. [10] The finite element method is effective for examining both the static deformation and vibration behavior of a structure and verifying a blade design. Modal testing is an experimental technique for obtaining the natural frequency of a structure. [11]

In the previous study, several designs with the different configurations of the hinged structure and rods were analyzed mainly regarding their load bearing capability and modal frequencies. Two schemes performed well in terms of the reduced moment and the increased frequencies. [12] The configuration of these two schemes was shown in Figure 1. [12] However, the results were not validated by the experiments and the rods' influencing mechanism on the modal frequencies was not clearly demonstrated. Herein, the study will focus on the two kinds of rotors' dynamic behavior and the influencing factors. Based on both the experiments and numerical study, the paper will show the discipline of the modal frequency and the damping with the change of the configuration of the rods.

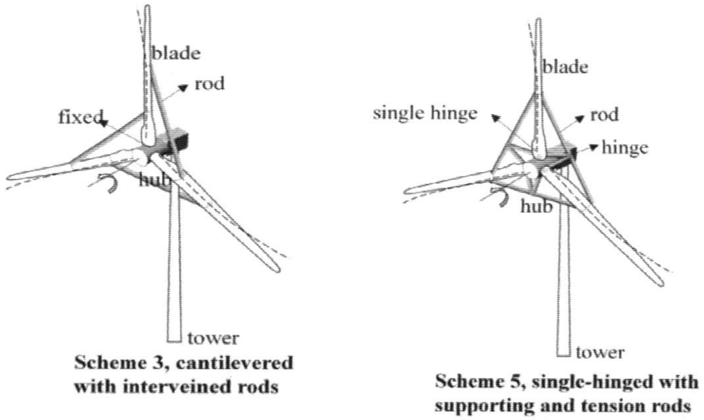

Figure 1. Illustrations of two novel wind turbine rotors, scheme 3 and scheme 5. [12]

2. Modal testing experiments

Modal testing experiments were completed on a platform developed in-house (Figure 2) to examine the rod's influence on the stiffness of the two blades [13]. The length of the tested blade was 1.35 m. The blade was made of fiberglass reinforced plastics and weighted 2.8 kg. The chord distribution of the blade was shown in Figure 2(c). To obtain clear and reliable signals, two INV 9823 acceleration sensors were mounted on gauge points close to the blade tip to record the blade response. The sensors and hammers were directly connected with an eight-channel acquisition card. Thirteen gauge points were stroked by the hammer in the chord-wise direction in order, and the excitation force and free vibration response were recorded. Data acquisition and signal processing (DASP) method was adopted to extract the modal frequency, damping ratio and mode shape. Each rod, made of aluminium alloy, was composed of two parts. The two separated parts were connected by threads. In the tested model, the rods connected with blade at No.6 gauged point, where is 47 % span along the blade length.

The modal frequency, damping ratio and mode shape of the cantilevered blade with and without rods were obtained in modal testing (Figure 3). Because the two acceleration sensors were mounted on the blade perpendicularly, vibration information in both the flap and edgewise directions were acquired (f: modal frequency in these two directions, N_m: mode number).

Because of the low stiffness in the flap direction compared with the edgewise direction, the lower modes were mostly flap modes; the third mode exhibited edgewise vibration. After the hinged rods were mounted in the rotating plane of the blade (Figure 2), the edgewise frequency rose significantly (Figure 3(a)). The hinged rod improves the in-plane stiffness of the blade. The frequency of the flap modes did not increase after the hinged rods were introduced. Each group of modal testing experiments was conducted three times, and the results seemed similar. The reduced frequency of the higher modes may be related with the slight change in stiffness due to the in-plane rods.

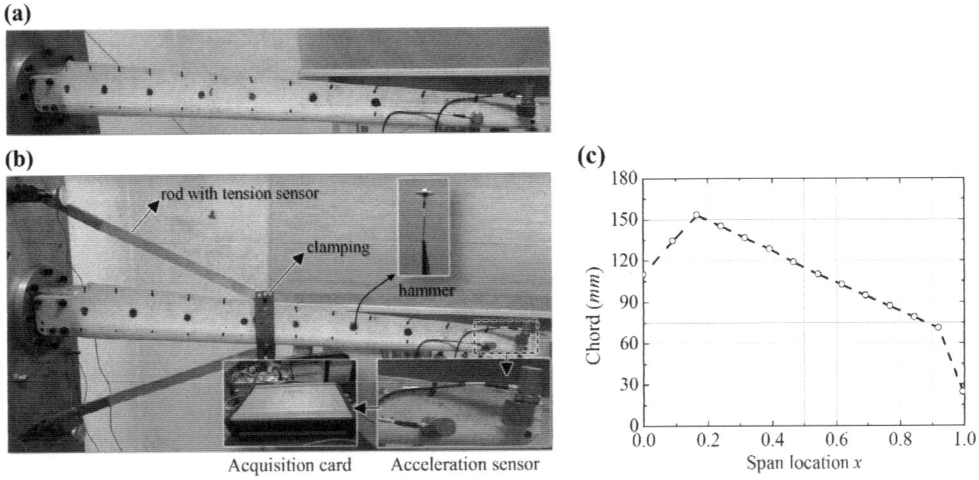

Figure 2. Modal testing experiments: (a) cantilevered blade; (b) cantilevered blade with hinged rods; (c) chord distribution of the tested blade.

Figure 3(b) illustrates the mode shapes and corresponding modal damping of the two blades in Figure 2. The upper part of Figure 3(b) shows the previous six modes for the blade with the hinged rods, while the lower part of Figure 3(b) shows the results of the cantilevered blade. The two blades had the same mode shapes for the previous four modes. However, because the rod added a DOF to the system, the blade with the in-plane rods had one more edgewise mode. Furthermore, introducing the hinged rods reduced the frequency of the higher modes: the flap and torsion coupling modes (sixth mode in upper Figure 3(b) and fifth mode in lower Figure 3(b). The hinged rods elevated the system's damping ratio, especially for the edgewise mode. This improved the reliability of the blade because a large structural damping ratio helps with the vibration attenuation and reduces the blade instability. The experiments demonstrated that the hinged structure and the rods effectively improve the blade's mechanical performance. Further study was needed to reveal the mechanism of the rods' influences on the modal frequencies of the rotor.

3. Factors influencing the rotor stiffness.

Different rod configurations enhanced the stiffness in schemes 3 and 5 of Figure 1. [12] To study the underlying principles, two dimensionless parameters were defined. For scheme 3, the normalized distances between the two rod locations and the blade root over the blade length were defined as x_1 and x_2 (Figure 4(a)). The three values of x_1 were kept constant, and x_2 was moved along the blade to obtain its influence (Figure 4(b) and (c)). The collective mode frequency f_{e1} and whirling mode frequencies $f_{e2,3}$ were normalized by the first edgewise mode frequency f_{e0}.

The interveined rods clearly had different effects on the collective mode and whirling modes. Looking at the curves' valleys in Figure 4(b), when $x_2 = x_1$, a local minimum of f_{e1} was obtained. When $x_1 = 0.1659$, the rotor stiffness could not be improved when $x_2 > 0.7187$. When $x_1 = 0.3285$, the rotor stiffness could not be improved when $x_2 > 0.6537$. The optimal location of x_2 to reach the maximum frequency was about 0.6 ~ 0.7 of the distance to the root. When $x_1 = 0.4586$, the optimal location of x_2 was about 0.1667. This could be derived from the curve's peaks. When $x_1 = 0.4586$, mounting the other rod end at locations away from $x_2 > 0.3285$ could not improve f_{e1}. These characteristics are distinct from the other two cases. When x_1 was close to the root, the curves had two peak values (Figure 4(b)).

Figure 4(c) shows the influence of the interveined rods on the second and third edgewise modal frequencies. The three curves with different values of x_1 had similar trends. When the rod location x_2

was moved forward to the blade tip, $f_{e2,3}$ increased significantly and then decreased. The frequency was improved the most when x_2 = 0.6–0.7. When x_1= 0.4586, $f_{e2,3}$ was markedly improved. Thus, the whirling modes were more affected by the interveined rods than the collective mode.

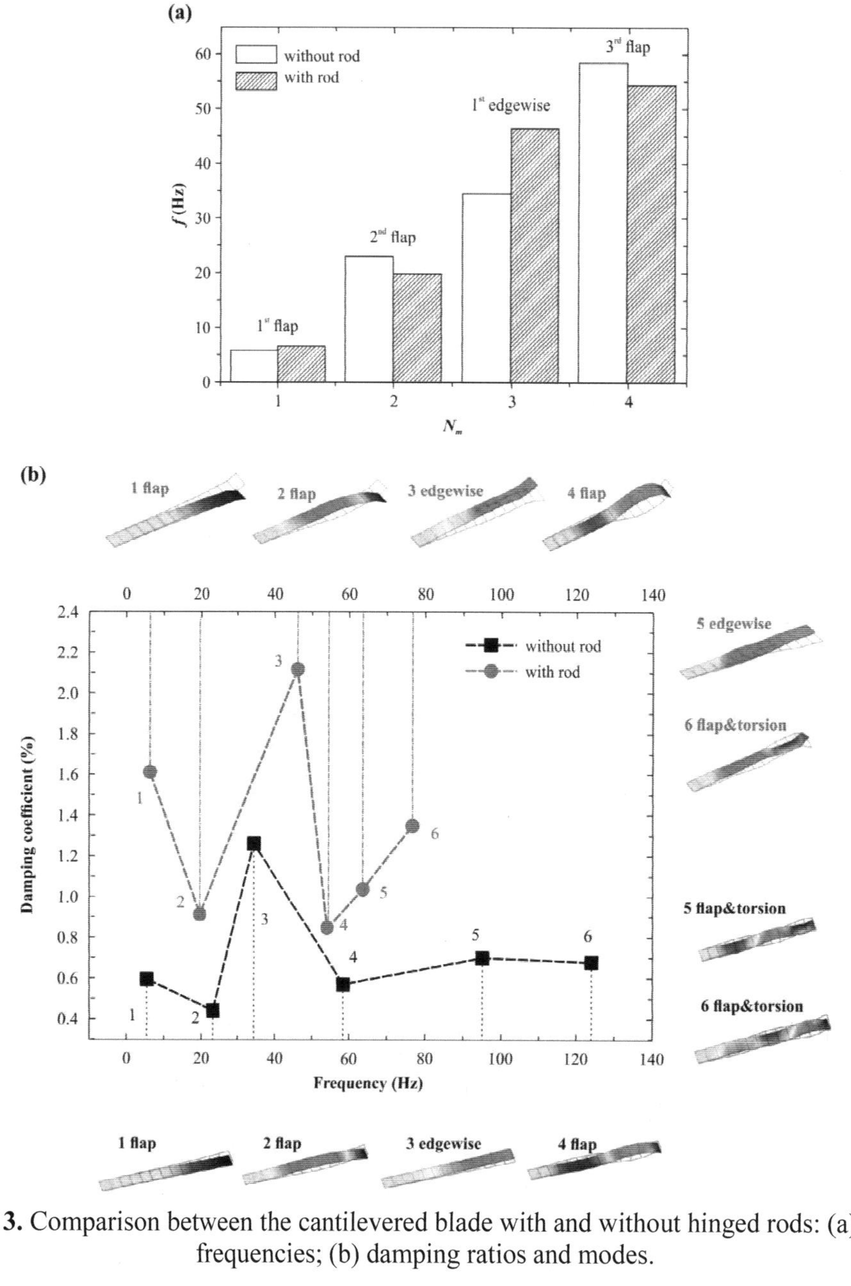

Figure 3. Comparison between the cantilevered blade with and without hinged rods: (a) modal frequencies; (b) damping ratios and modes.

In scheme 5, the rod configuration affected the frequency change in a different pattern (Figure 5). Two parameters were defined to describe the locations of the supporting rods and hinged rods: x denotes the normalized rod-mounting location on the blade, while R denotes the distance of the joint of the supporting rod and the tension rod away from the hub center. R was divided by the hub radius R_0 to get the dimensionless parameter R/R_0.

Figure 5(b) shows the influence of the rod-mounting location on the frequencies of the first edgewise mode group. The whirling mode frequencies were clearly improved more than those of the collective mode. A different influence principle was found. When R/R_0 was set to a small value (e.g., 2.5), f_{e1} slightly increased. However, it decreased as the joint location R/R_0 moved away from the hub. When $x = 0.1659$ and 0.3285, once $R/R_0 > 2$, f_{e1} could be improved. However, for the whirling modes, when the joint location moved in the range of [2, 5], the frequencies increased quite quickly until they reached a steady value.

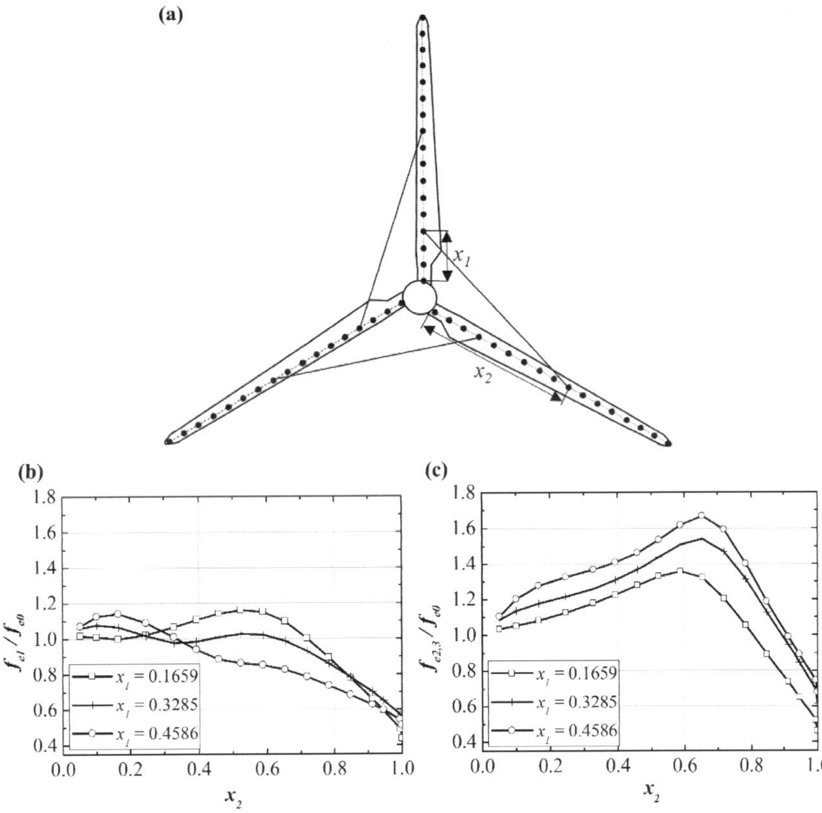

Figure 4. Influence of the locations of the interveined rods on the modal frequency of the rotor: (a) defined parameters; (b) frequency of the collective mode; (c) frequency of the whirling modes.

Figure 5(c) summarizes the pattern of how the tension rods' location affecting the frequencies. Both of the frequencies of the collective mode and whirling modes first increased and then decreased. This impact was more obvious for the whirling modes than the collective mode. f_{e1} was most improved when the tension rod was located at $x \sim 0.6$ from the blade root. $f_{e2,3}$ were most improved when the tension rod was located at $x \sim 0.7$ from the blade root. As shown in Figure 5 (c), when the tension rod was located too far from the blade root (e.g., $R/R_0 \sim 10.25$), f_{e1} showed almost no improvement.

4. Conclusions

In this study, the dynamic behavior of the cantilevered blades with the interveined rods and the single hinged blades with the bisymmetrical rods were investigated. The conclusions were drawn as follows:

1) The modal testing results showed the potential benefits of the hinged rods to enhancing both the blade stiffness and damping of the edgewise modes.
2) The dimensionless parameters x_1, x_2, x and R/R_0 were defined to reveal the principles of the edgewise modal frequencies for scheme 3 and 5. The influencing curves of the rods' location on the frequencies showed large differences with the maximum frequency occurring at the different locations.

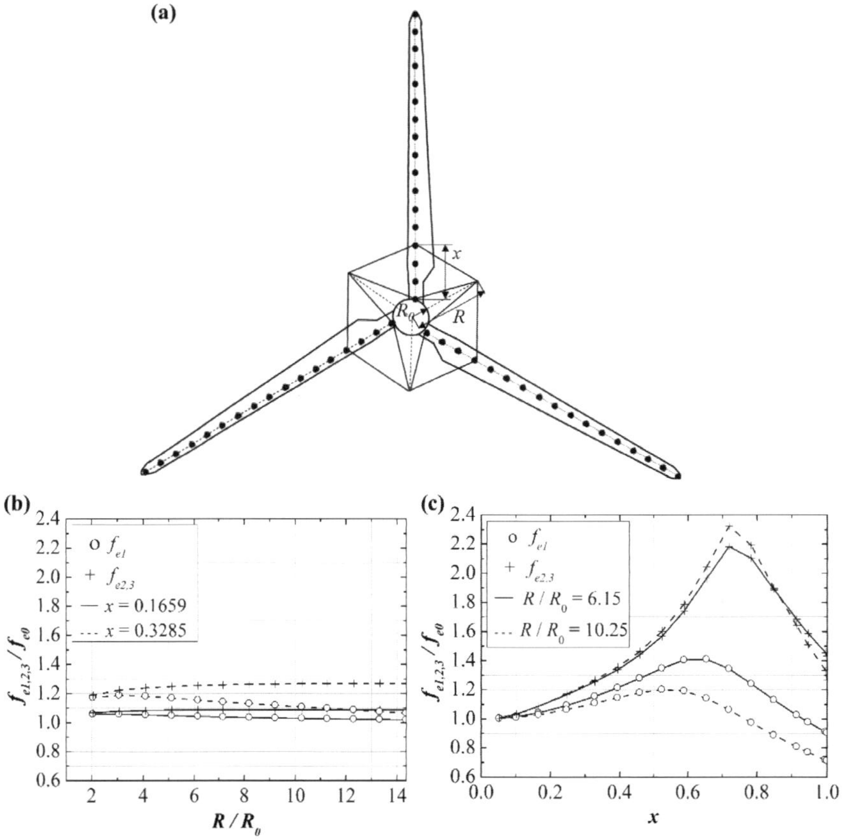

Figure 5. Influences of the locations of the bisymmetrical rods on the edgewise frequency of the rotor: (a) defined factors; (b) the hinged locations' influences; (c) the tension rods' location's influences.

Acknowledgments

The study was supported by the National Natural Science Foundation of China (No. 51575296). The authors gratefully acknowledge this support.

References

[1] R McKenna, V. D. Leye P O, W Fichtner, Renew Sust Energ Rev 2016; 53:1212-1221.
[2] V A Riziotis, S G Voutsinas, E S Politis, et al., Wind Energy, 2004; 7(4):373-392.
[3] P K Chaviaropoulos, E S Politis, D J Lekou, et al., Wind Energy 2006; 9(1-2):163-177.
[4] T. Burton, D. Sharpe, N. Jenkins and E. Bossanyi, Wind energy handbook (John Wiley & Sons, 2001).
[5] J. P. Baker, C. P. Van Dam and B. L. Gilbert, Flatback airfoil wind tunnel experiment. Report, Sandia National Laboratories, Albuquerque, New Mexico, 2008.
[6] Roth-Johnson, R. E. Wirz and E. Lin, Renew Energy 2014; 71: 133–55.
[7] P. Roth-Johnson and R. E. Wirz, Wind Energy 2014; 17(3): 397–411.
[8] K. Y. Maalawi and H. M. Negm, J Wind Eng Ind Aerodyn 2002; 90(8): 961–86.
[9] R. W. Vesel and J. J. McNamara, Renew Energy 2014; 66: 391–401.
[10] M. Tarfaoui, H. Khadimallah, O. Shah and J. Y. Pradillon, Fourth international conference on power engineering, energy and electrical drives (PowerEng), Istanbul, Turkey, 2013.
[11] H. Lu, P. Zeng, L. Lei, ICMAE 2017 (Accepted) 2017.
[12] F. M. Jensen, B. G. Falzon, J. Ankersen and H. Stang, Compos Struct 2006; 76(1): 52–61.
[13] H. Lu, P. Zeng, L. Lei, Int J Energ Res, 2016, 40(13): 1773-1787.

Research on Standard and Automatic Judgment of Press-fit Curve of Locomotive Wheel-set Based on AAR Standard

Jun LU[1], Jun XIAO[1], Dong Jun GAO[1], Shu Yu ZONG[1] and Zhu LI[1]

[1] School of Mechanical and Electrical Engineering, Wuhan University of Technology, Wuhan, Hubei ,430070, China

Email: jun10099@163.com

Abstract. In the production of the Association of American Railroads (AAR) locomotive wheel-set, the press-fit curve is the most important basis for the reliability of wheel-set assembly. In the past, Most of production enterprises mainly use artificial detection methods to determine the quality of assembly. There are cases of miscarriage of justice appear. For this reason, the research on the standard is carried out. And the automatic judgment of press-fit curve is analysed and designed, so as to provide guidance for the locomotive wheel-set production based on AAR standard.

1. Introduction

The wheel-set is core parts of locomotive bogie. Its assembly method is keyless fitting, that is, the wheel-set assembly equipment is used to combine the wheel and axle assembly. [1, 2] In the AAR standard, the wheel-set is assembled by cold-pressing. [3] As the wheel-set is directly contacted with railway, its assembly quality is greatly related to the running safety of locomotives and vehicles. In the process of production, the real-time press-fit curve formed by NC equipment to record the press-fit force and displacement, which is the important criterion to judge whether the quality is qualified. And it also the only standard to evaluate the quality of wheel-set assembly. [4]

In the Association of American Railroads Standards and Recommended Practices Manual - Wheels and Axles, the press-fit curves are specified and the qualified and unqualified curves are enumerated. Many defects description of press-fit curve is provided in great detail. Based on standard, it is beneficial to standardize the production and improve the quality of wheel-set.

In this paper we put forward an automatic mathematical model by analysis about press-fit curve. And then the corresponding computer automatic judgment program is compiled. It can provide reference for the optimization of the production and maintenance standards on wheel-set. It also helps to the design of the automatic judgment program based on other standard. [5, 6]

In the AAR standard, the press-fit curve of X axis is specified as force F, and Y is specified as displacement L. It's shown in Figure 1. However, in the actual production, wheel-set assembly machine is often setting displacement as the X axis, and the force is Y axis. This can reflect the tendency of the force to change with the displacement so that the staff can better monitor the press-fit force according to the press-fit curve. [2, 4]

IOP Conf. Series: Materials Science and Engineering **326** (2018) 012010 doi:10.1088/1757-899X/326/1/012010

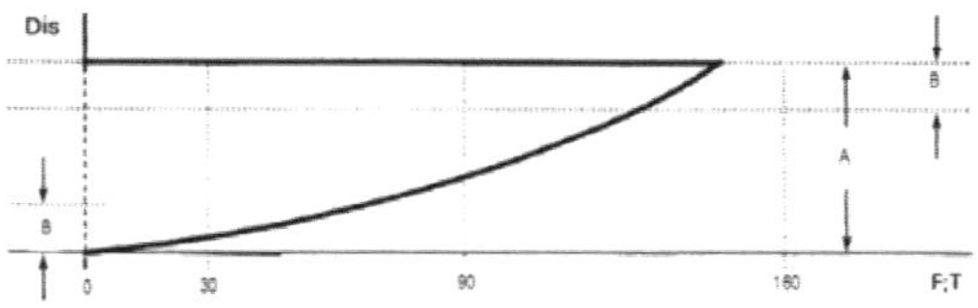

Figure 1. The press-fit curve of AAR standard

2. AAR standard analyses

When the wheel-set is pressed, the shape of press-fit curve and the final press-fit force are mainly related to the factors such as the amount of interference, the axle taper, the surface, the roughness, the lubrication and the material of axle. Theoretically, the length of AAR standard press-fit curve can refer to the China railway standard, which is indicated as follows: [7]

$$L = (S + A - K - R) \tag{1}$$

Among them:
L—The length of press-fit curve.
S—The length of wheel hub bore.
A—The length of the wheel seat extends from the outer end of the wheel hub bore.
K—Taper inlet length of front end of wheel seat.
R—Inner arc radius of wheel hub.

According to standard the length of press-fit curve is divided into three parts in the automatic judgment process. [8] It should be divided into 0 to 25% to 75% to 100% of the displacement. The displacement segment according to the total number of point n divided:

a. The starting section: In the range of (0~25) % of the displacement, the point number:

$$i = 1\sim0.25n \tag{2}$$

b. The middle section: In the range of (25~75) % of the displacement, the point number:

$$i = 0.25n\sim0.75n \tag{3}$$

c. The end section: In the range of (75~100) % of the displacement, the point number:

$$i = 0.75n\sim n \tag{4}$$

2.1. Wire frame judgment

The standard points out that “The displacement, maximum value, minimum value of tonnage should be in accordance with AAR MSRP Section G-II Wheel and Axle Manual. At the beginning of press-fit, the equipment generates the judgment line of the press mounting curve automatically.” When any part of the press-fit curve exceeds the outside of judgment dashed line, it can be judged that the wheel-set is unqualified.

2.2. Force rise sharply

P The standard points out that "The 'entry spike' at the beginning of press-fit must not exceed 30 ton or three times the tonnage after the drop-off whichever is less. For example, if the spike is no greater than 30 ton and the tonnage after the drop-off is no less than 10 ton, the press-fit is acceptable.

The surface of wheel and axle, such as the uneven surface or the wheel and the axle is not moderate which will cause the starting point force to rise sharply. Therefore, there is a strict requirement for the force at the starting point, as follows:

$$T_1, T_2, T_3 < 30; \quad if\ T_m < T_{m-k},\ then\ \frac{T_{m-k}}{T_m} < 3 \tag{5}$$

Among them:
m<0.25n, n is the total number of discrete points of the press-fit curve.

2.3. Determination of the maximum force
The standard indicates: "If the maximum force on press-fit curve is greater than the maximum allow value, the curve is not qualified. This requires all points of the curve to be within the maximum allow value." This regulation is similar to 2.1:

$$T_i \le T2max, i \in \{0,1,2,3 \dots n\} \tag{6}$$

2.4. Determination of the final force
The standard reads: "If the final press-fit force is less than the minimum allow value, the curve is judged to be substandard". This provision is implied in the 2.1-line frame judgment, the requirement:

$$T4 > T2min \tag{7}$$

2.5. Position judgment
The standard points out: "The press-fit curve is not allowed to fall within the range of 2 tons of the maximum allow force or the minimum allow force. Otherwise it is judged to be unqualified." This means that the last force must be less than the maximum force of 2 tons, and higher than the minimum force of 2 tons"

$$T2min + 2 < T4 < T2max - 2 \tag{8}$$

2.6. Force steady rise
The standard points out: The force during the press-fit shall be rising uniformly, and the tonnage shall be established continuously. The 2.6 and 2.7 shall be regarded as coincidence. And the force shall be maintained even rise in the case of no force decline.

2.7. Force down
The standard says: "Starting from the first drop point of the first 25% displacements (no matter how much the force is reduced) until the 75% displacement is calculated. Or if there is no drop in the first 25% displacement, the curve is automatically judged from the first displacement point until the 75% displacement. According to the AAR standard, the force drop before and after the data (including the continuous force decrease of multiple points) exceeds 2 tons and the quality is judged to be unqualified. When the geometry of the press-fit faces is defective, the force at a certain position will decrease briefly. If the lubrication film is not adequately coated, it will cause the force to rise up somewhere and then decrease in tons. These conditions have great influence on the quality of press-fit. Therefore, the curve is required to meet the following conditions:

$$if\ T_i < T_{i-1},\ T_{i+1} < T_i,\ \dots,\ T_{i+k} < T_{i+k-1};$$

$$then\ T_{i-1} - T_{i-k} < 2 \tag{9}$$

2.8. Curve completely flat or force drop only
The standard points out: in the first 75% of the press-fit curve, the maximum force cannot appear. That means that the maximum force is at the end section, otherwise the curve is unqualified.

$$\text{if } T_i = T2, \text{ then } i > 0.75n \tag{10}$$

2.9. Force fluctuation

The standard states: judging from the start at the 75% point of the press-fit curve. If there is one or more times more than 2 tons of tons (or cumulative drop tons), then determine the end of the press-fit curve appears fluctuation. At the same time, if the end of press-fit force is below the minimum tonnage value, the curve is judged to be unqualified.

$$if\ \ T_i < T_{i-1},\ T_{i+1} < T_i,\ \ \ldots,\ T_{i+k} < T_{i+k-1},\ then\ T_{i-1} - T_{i-k} < 2,\ T_i \geq T2_{min} \quad (11)$$

2.10. Curve is not concise

The standard states: if the press force does not meet the minimum press force requirements before the last five acquisition points of the press curve, even if the force for any of the five acquisition points is higher than the minimum tonnage, the curve is still judged to be unqualified.

$$exist\ T_i \geq T2\ min\ \ ,\ and\ \ i < n - 5 \quad (12)$$

2.11. Curve reverse crossing

The standard indicates: "when the pressing curve is more than 30 tons of the pressing force, if there is a press fit force, and then return to the limit of 30 tons (including 1 tons of down tons), the curve is judged to be unqualified". It is pointed out in the standard: "when the pressing curve appears first to exceed 30 tons of pressing force and then goes back to the limit of 30 tons (including 1 tons of force down), the curve is judged to be unqualified." This regulation can prevent the interference of the locomotive due to the interference fit is not firm.

$$if\ \ T_i \geq 30,\ \ then\ \ T_{i+k} \geq 30 \quad (13)$$

$$if\ \ T_i \geq T2\ min\ \ ,\ \ \ then\ T_{i+k} \geq T2\ min \quad (k=1,2,3\cdots) \quad (14)$$

3. Automatic Judgments Mathematical Model

The basis of establishing the mathematical model of automatic judgment is based on the press-fit curve and the rules in the standard. In the AAR standard, it is necessary to establish the judgment line frame of the standard curve, and set the boundary condition according to the given parameter or the parameter of the calculation formula. In the AAR standard, the numerical value of press-fit force is expressed as T units, and the press-fit curve records are often used KN units, so the corresponding parameters should be adjusted according to the actual situation. As for the parameters in the standard, they are expressed in figures. And the graphics of coarse images can be expressed in the mathematical form and quantified, and then compiled into an automatic judgment program. According to the standards, we design the following model in Fig.2.

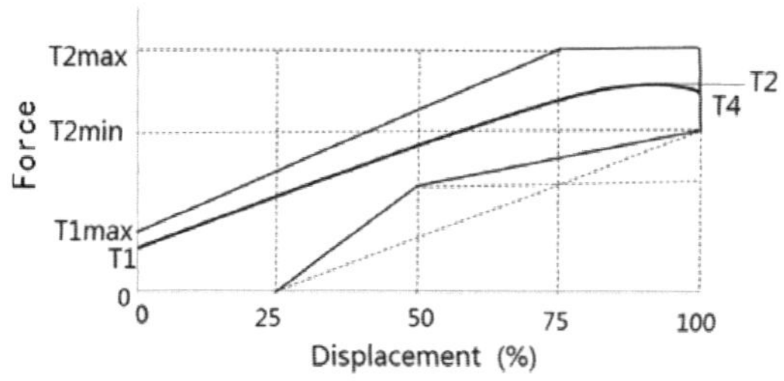

Figure 2. Curve judgment frame

Among them:

T1max- The maximum allow force at the starting point of curve
T1-The actual force at the starting point of the force curve
T2max- The maximum allow force
T2min- The minimum allow force
T2- The maximum force in a curve
T3- The down force at the end of the curve (T2-T4)

IOP Conf. Series: Materials Science and Engineering **326** (2018) 012010 doi:10.1088/1757-899X/326/1/012010

T4- The last point force in a curve

4. Test and analysis of results

The computer program is compiled according to the automatic judgment model. It is possible to realize the automatic judgment of a curve theoretically. But its reliability needs to be verified by tests. The control system of NC wheel-set assembly machine was designed and developed in the C + + Builder platform. Through the above research and analysis, we are write programs on the basis of the mathematical model. Through the test and historical production data import program to judge. We can verify the accuracy that model identify defects and fit the standard. [9] The automatic judgment process is showed in Fig.3.

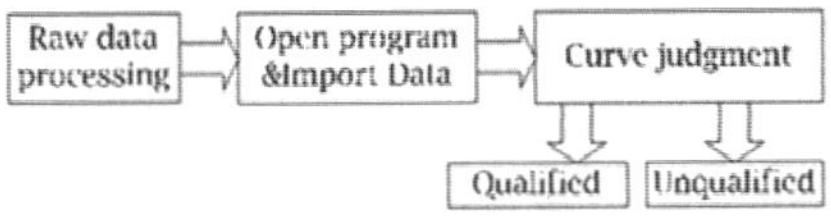

Figure 3. The process of automatic judgment

In the production of wheel-set assembly plant, the majority of qualified curves are shown in Fig.4. When there is a force-drop on the right wheel in Fig.5, the defect cannot be accurately identified only by the human eye. The last force exceeds the T2max in Fig.6. The program has a good judgment result. It can be seen that the accuracy is not high by manual judgment. The algorithm is more effective to solve this problem.

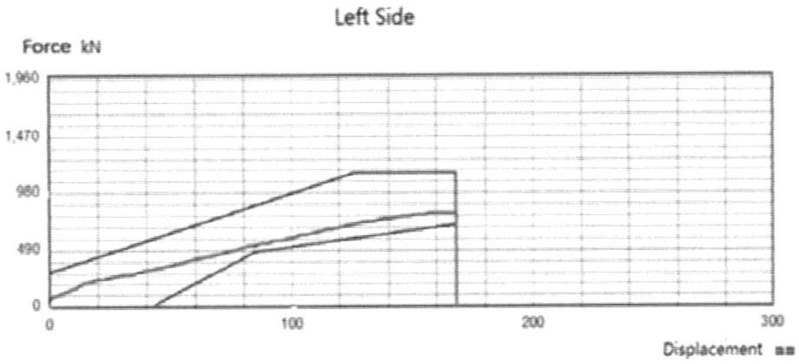

Figure 4. Qualified curve

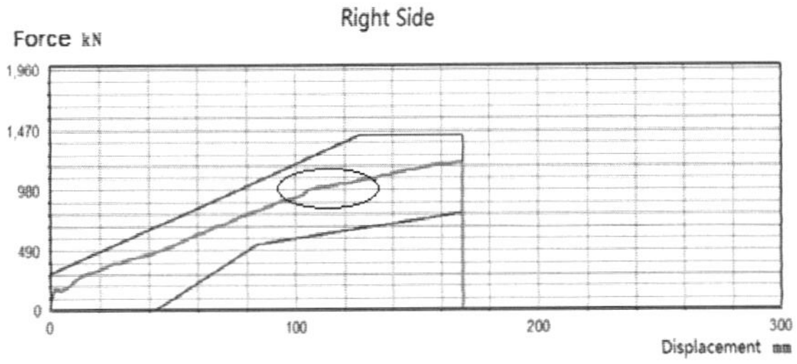

Figure 5. Unqualified curve 1

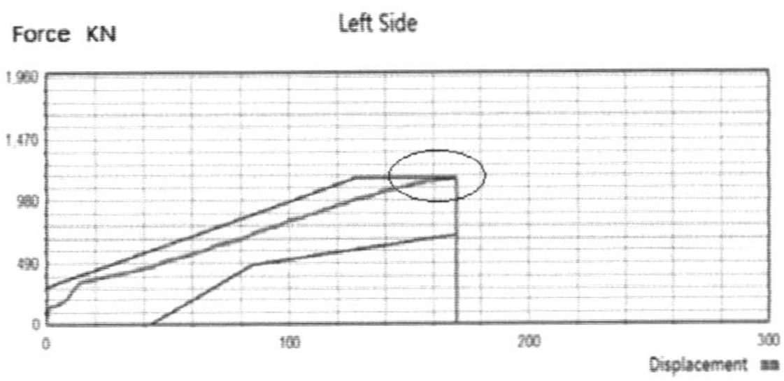

Figure 6. Unqualified curve 2

5. Conclusions

In this paper, we study the technical requirements of wheel-set assembly in AAR standard carefully. We have studied the judgment algorithm, and construct the mathematical model of automatic judgment. In fact, qualities testing artificially and automatically are required to be carried at the same time. As the curve that is very difficult for workers to judge, the automatic judgment program can make very good effect. Statistical data show that the accuracy of the automatic judgment program is approximately 100%. The test results verify the effectiveness of the research. The model greatly avoids the mistakes of previous manual judgment and improves reliability of wheel-set. It has important application value for ensuring safety of railway vehicles.

References

[1] AAR MSRP Section G-II Wheel and Axle Manual. Association of American Railroads.2004
[2] Han Jianbin. Research on wheelset press curve recording and automatic judgement system [D]. Wuhan University of Technology, 2012.
[3] Feng Shaoyan. Comparative study on standard of wheelset assembly [J]. Locomotive and rolling stock technology, 2014 (2): 25-26.
[4] Du Hongguang, Jiang Bin, Pei Peng, et al. Mathematical Model for Automatic Judgement of Wheelset Press and Put Curves[J]. Urban Mass Transit, 2014, 17 (9): 26-28.
[5] Tian Baoshuan. The Analysis and Discussion of Strength Design and Testing Standards for Railway Freight Carbodies in Our Country and Abroad (To Be Continued) [J]. ROLLING STOCK, 2009, 47 (6): 26-30.
[6] Huang Yinxia. On the US railway technical regulations and standards [J]. China Railway, 2006 (12): 33-35.
[7] Chen Yanling, Pan Hongsheng, Wang Lizhi. Effect of the Wheel-set Protruding Assembly on the Press Curve of Press Mounting [J]. ROLLING STOCK, 2008, 46(7): 32-34.
[8] Jing Biao. Analysis and suggestion on the roughness of wheel hub and wheel hub bore of wheelset of EMU [J]. ROLLING STOCK, 2013, 51 (3): 42-43.
[9] Xu Li, Liu Manhua. Analysis of the EN standards on press-fit of wheelset and the determination of curves for press-fit[J]. LOCOMOTIVE & ROLLING STOCK TECHNOLOGY, 2009(3):27-29.

IOP Conf. Series: Materials Science and Engineering **326** (2018) 012011 doi:10.1088/1757-899X/326/1/012011

Federal Aviation Regulations – National Aviation Regulations of Russia

O Chernykh[1, a] and M Bakiiev[2, b]

[1]Department of Civil Aviation Engineering, College of Civil Aviation, Nanjing University of Aeronautics and Astronautics, 29 Jiangjun Avenue, Nanjing, 211106, Jiangsu province, China
[2]Department of Fuselage Airframe Computer Projects, Antonov State Company, 1 Tupolev Street, Kyiv, 03062, Ukraine

[a]alex@nuaa.edu.cn; [b]mambet12@gmail.com

Abstract. Chinese Aerospace Engineering is currently developing cooperation with Russia on a wide-body airplane project that has directed the work towards better understanding of Russian airworthiness management system. The paper introduces national Aviation regulations of Russia, presents a comparison of them with worldwide recognized regulations, and highlights typical differences. They have been found to be: two general types of regulations used in Russia (Aviation Regulations and Federal Aviation Regulations), non-unified structure of regulations on Aircraft Operation management, various separate agencies responsible for regulation issuance instead of one national aviation authority, typical confusions in references. The paper also gives a list of effective Russian Regulations of both types.

1. Introduction

Recently, Russian aviation regulations have become a challenge for Chinese professionals in Aerospace industry due to a reason of established cooperation between China and Russia in Aerospace matters. In particular, the work of Chinese COMAC Aircraft Company has currently been extended to a new long range wide-body commercial airplane (C929/C939) which has been agreed to be in cooperation with Russia.

China has long been an operator of aircraft purchased from abroad but now China has joined a narrow group of countries which are capable of designing and manufacturing their own aircraft. Now, the entire aircraft development process along with aircraft airworthiness substantiation has become of the highest importance. It's well known, an airplane cannot enter service unless a manufacturer has demonstrated that an airplane complies with applicable airworthiness regulations and a national aviation authority agrees they've been met. Since development process has involved Russia as the "third party", Russian airworthiness regulations have to be additionally taken into consideration. This fact has been put on the basis of further improvement of educational level in Chinese Aerospace-related universities [1].

However, the methodology of working with Russian regulations is quite different due to their dissimilarity from local Chinese Civil Aviation Regulations (CCAR) and worldwide known Federal Aviation Regulations (FAR or 14 CFR, being part of Title 14 of the Code of Federal Regulations) of the Federal Aviation Administration (FAA) of the U.S.A. or Certification Specifications (CS) of the European Aviation Safety Agency (EASA) of the European Union.

Russian aviation regulations include two principal types. They have been:
(1) Aviation Regulations and
(2) Federal Aviation Regulations.

The full list of Russian aviation regulations including all documents of the two above-mentioned types hasn't been found in open English-language sources and is not even provided in Russian-language informative space. The book on certification of aviation products written by Krasotkin [2] contains only the list of the 1st type, the Aviation Regulations, remarking, not covering all of them, and no any document of the 2nd type, the Federal Aviation Regulations. Surprisingly, even though the early Russian Federal Aviation Regulations have started to be issued since 1998 [3], they haven't been mentioned in the book of Krasotkin [2] published 9 years later in 2007.

However, the regulations can be found on a number of official websites of Russian authorities. The Aviation Regulations are provided on the website of the Interstate Aviation Committee (IAC) which is responsible for their issuance and handling [4]. Having been issued by various authorities, the Russian Federal Aviation Regulations have been disposed on websites of Russian ministries and agencies, but it has been made in an uncoordinated manner. Each authority's website only provides documents of own direct influence and control [1]. Among the mentioned, there are the following authorities which govern aviation industry in Russia: the Government of the Russian Federation [5], the Ministry of Transport [6], the Ministry of Defense [7], the Federal Air Transport Agency [8].

Unfortunately, there is no any full unified list of the Federal Aviation Regulations, being provided so far. It's worth noting, the first attempt to compose such a unified list of Russian aviation regulations has been made by the Federal State Unitary Enterprise Scientific Research Institute of Standardization and Unification in 2014 resulting in establishment of the Unified System of Aviation Standards [9], however its website hasn't been properly updated since 2014, so it has been found providing regulations which are no more effective. One of the main purposes for this paper has become presenting the full list of Russian aviation regulations collected by the authors as of November 2016.

2. Aviation Regulations (AP)

The Aviation Regulations (AP) [10] have been issued by the Interstate Aviation Committee (IAC) and are effective in all member countries of the interstate treaty. Russia, being a member, has also adopted this set of regulations. The Aviation Regulations are commonly shortened to a confusing abbreviation *"AP"* which has come from a transliteration *"Aviatsionnye Pravila"* of their Russian name *"Авиационные Правила (АП)"*.

The regulations have been issued in Russian language exceptionally [10]. The content has been arranged in Parts with a serial number *"AP-XXX"*. The list and titles of the Aviation Regulations (AP) can be found on the official website of the Interstate Aviation Committee (IAC) [4] and are given in Appendix A with explanations.

In general, the Parts of Aviation Regulations (AP) follow the typical content structure and serial numbers of worldwide commonly accepted structure of regulations (such as, of the American FAR regulations). This has been an outcome of continuous harmonization process started in 1990s by the Interstate Aviation Committee [2]. But it's worth noting that the Aviation Regulations (AP) do not have as many parts and do not cover as many topics as the American FAR regulations or European CS rules do. The issued Parts primarily cover Initial Airworthiness matters and are all displayed in this paper.

Moreover, it has recently been announced that the seven AP Parts (AP-23, AP-25, AP-27, AP-36, AP-139, AP-170 Volume I, and AP-183) have become ineffective in Russia and the relevant governmental orders have been signed [11, 12]. However, the Parts themselves are still effective as the Parts of the AP regulations managed by the Interstate Aviation Committee.

3. Federal Aviation Regulations (FAP)

The Federal Aviation Regulations (FAP) [3] have been issued as Orders of various Russian ministries and agencies [1, 5-8]. Their abbreviation is shortened as *"FAP"* which has also come from a transliteration *"Federal'nye Aviatsionnye Pravila"* of the Russian name *"Федеральные Авиационные Правила (ФАП)"*. The FAP regulations are considered Russian national regulations being effective in Russia exceptionally.

Such a confusing name of Russian regulations sounding completely identical to the American Federal Aviation Regulations (FAR) has been grounded on the fact that the full name of Russia has been the Russian Federation. All Russian governmental agencies hold the Federal status, that is why the word *"Federal"* has also come into the name of Russian national regulations. To avoid misunderstanding, the abbreviation *"FAP"* has been established.

3.1. The detailed list of the FAP regulations

The FAP regulations have also been issued in Russian language only [3]. The content has been distributed among numerous Orders with a serial number and a year of issuance, such as *"Order XXX (Year)"*. The full list of the FAP regulations hasn't been found in open sources. Having been collected by the authors, they are listed in Appendix B according to responsible authorities.

Studying the FAP regulations and their arrangement of serial numbers for any similarity with the American FAA, European EASA regulations, and regulatory material of the International Civil Aviation Organization (ICAO), no correspondence has been found. For instance, "*Order 147 (2008).* Requirements to air crew men, aircraft maintenance engineers, and flight support crew / flight dispatchers in Civil Aviation" [13] covers similar matters as the following American FAA regulations [14]:

- FAR Part 61 – Certification: Pilots, flight instructors, and ground instructors,
- FAR Part 63 – Certification: Flight crewmembers other than pilots,
- FAR Part 65 – Certification: Airmen other than flight crewmembers;

or European EASA regulations – CS parts, including Acceptable Means of Compliance (AMC) [15]:

- CS-FCD – Flight Crew Data,
- AMC Part-ARA – Authority Requirements for Aircrew,
- AMC Part-ORA – Organisation Requirements for Aircrew,
- AMC Part-CC – Cabin Crew,
- AMC Part-FCL – Flight Crew Licensing;

or regulatory material of ICAO [16]:

- Annex 1 – Personnel Licensing,
- Doc 9868 – Personnel Training.

3.2. "FAP-XXX" reference using a serial number of a corresponding American FAR regulation

However, some of the Orders, in addition to their serial number and a year, have been commonly referred to as *"FAP-XXX"* with a separate serial number as it has been noted in the list of the FAP regulations for positions 55-61 (see Appendix B). These "additional" serial numbers are different from the serial numbers of the relevant Orders. But the "additional" numbers have been found to exactly correspond to the relevant parts of the American FAR regulations [14]. For instance:

- "FAP-67". *Order 50 (2002) of the Ministry of Transport.* Medical examination of air crew, dispatchers, flight attendants, pilot students, and entrants to schools of Civil Aviation.

>>> FAR Part 67 – Medical standards and certification.

- "FAP-119". *Order 246 (2015) of the Ministry of Transport.* Requirements to juridical persons, private entrepreneurs, who implement commercial air transportation.

>>> FAR Part 119 – Certification: Air carriers and commercial operators.

- "FAP-142". *Order 289 (2015) of the Ministry of Transport.* Requirements to educational institutions and organizations preparing aviation specialists of proper level.

>>> FAR Part 142 – Training centers.

- "FAP-145". *Order 285 (2015) of the Ministry of Transport.* Requirements to juridical persons, private entrepreneurs, who implement maintenance of Civil Aviation aircraft.

>>> FAR Part 145 – Repair Stations.

No correspondence between these "additional" *"FAP-XXX"* numbers and the European EASA or ICAO regulation denominations [15, 16] has been found.

But it's worth noting that neither of above-mentioned serial numbers *"FAP-XXX"* can be found in body of a relevant Order, for example [17]. It means that these references *"FAP-XXX"* can be treated as a professional slang intended to simplify work with the FAP regulations just indicating the corresponding American FAR regulations but NOT having any legal power.

3.3. "FAP-XXX" reference using a serial number of an Order

On the contrary, it has also been found a case [18] that those "additional" designations of *"FAP-XXX"* sometimes use the very serial numbers of Orders losing a year reference. For instance, *"Order 216 (2009)"* can be referred to as *"FAP-216"*. The lack of year of issuance may bring significant confusion as Orders are issued according to their series each year independently. So they are different for various years, such as the "*Order 216 (2009).* Requirements to air traffic dispatchers and parachute instructors" and the "*Order 216 (2015).* Requirements to juridical persons who implement aeronautical service for aircraft of Russian Federation airspace users" both obviously have various contents.

As a result, it is advised to avoid any reference like *"FAP-XXX"* instead of legally supported Order's serial number with a year of issuance like *"Order XXX (Year)"*. Otherwise, it may bring unreasonable confusion into work with Orders.

In general, the Russian Federal Aviation Regulations (FAP) are mainly used for Aircraft Operation and Continued Airworthiness management. But the structure of documents significantly differs from American or European regulations as they display their own content structure [3,13,17] not following typical operation regulations such as the American FAR parts [14]:

- FAR Part 91 – General operating and flight rules,
- FAR Part 121 – Domestic, flag, and supplemental operations,
- FAR Part 135 – Commuter and on demand operations;

Or the European CS parts [15]:

- CS-AWO – All Weather Operations,
- AMC Part-ARO – Authority Requirements for Air Operations,
- AMC Part-ORO – Organisation Requirements for Air Operations,
- AMC Part-CAT – Commercial Air Transport Operations;

And being very different from ICAO regulatory documents [16]:

- Annex 6 – Operation of Aircraft,
- Doc 8168 – Aircraft Operations.

The absolutely non-unified structure of the Russian Federal Aviation Regulations (FAP) is assumed to be a result of distribution of Airworthiness management responsibilities among separate agencies [1] instead of one national aviation authority that has been common for other countries.

4. Conclusion

The work with Russian regulation management system has been found to have certain difficulties for Chinese Aerospace industry professionals not just because of regulation content is arranged into parts differently from Chinese regulations. The Russian regulations have not been fully harmonized with unified American or European Union regulations either, utilizing two different types of regulations.

While the aspects of Initial Airworthiness have been concentrated in the 1st type – Aviation Regulations (AP), which parts correspond to the worldwide commonly used structure of regulations. Aircraft Operation and Continued Airworthiness matters have been covered in the 2nd type – Federal Aviation Regulations (FAP) which have been treated as national Russian regulations. However, the FAP regulations follow their own arrangement of contents distributed among numerous Orders which have been issued and managed by a number of Russian authorities with their own field of influence.

Some confusing key dangers have also been highlighted to attract attention of professionals and minimize the risk of misunderstanding the Russian Federal Aviation Regulations (FAP), such as referring to Orders with an indicating reference of *"FAP-XXX"*. It has been advised to avoid such cases referring to as *"Order XXX (Year)"* exceptionally.

The above-mentioned aspects have been put the highest concern for the further development of international relations in Aerospace matters.

Appendix A

The full list and titles of the Aviation Regulations (AP) issued by the Interstate Aviation Committee (IAC) (the IAC website's translation into English [4]):

1. AP-21. Certification procedures for Aeronautical Products, Design Organisations and Manufacturers.
2. AP-23. Airworthiness standards for Light Airplanes. *(invalid in Russia)*
3. AP-25. Airworthiness standards for Transport category Airplanes. *(invalid in Russia)*
4. AP-OLS. Airworthiness standards for Very Light Airplanes. *(AP-VLA – literally)*
5. AP-27. Airworthiness standards for Normal category Rotorcraft. *(invalid in Russia)*
6. AP-29. Airworthiness standards for Transport category Rotorcraft.
7. AP-31. Airworthiness standards for Manned Free Balloons.
8. AP-33. Airworthiness standards for Engines.
9. AP-34. Emission Requirements.
10. AP-VD. Airworthiness standards for Auxiliary Power Units. *(AP-APU – literally)*
11. AP-35. Airworthiness standards for Propellers.
12. AP-36. Noise Certification. *(invalid in Russia)*
13. AP-39. Airworthiness Directives.
14. AP-139. Certification of Aerodromes. *(invalid in Russia)*
    - Volume I – Procedures for Certification of Aerodromes.
    - Volume II – Certification Requirements to Aerodromes.
15. AP-145. Repair Stations.
16. AP-170. Certification of Aerodrome and En-Route Equipment.
    - Volume I – Procedures for Certification of Aerodrome and En-Route Equipment. *(invalid in Russia)*
    - Volume II – Certification Requirements to Aerodrome and En-Route Equipment.
17. AP-183. Aviation Register Representatives. *(invalid in Russia)*

One more example of a transliterated abbreviation found in the list needs to be explained:

- AP-OLS stands for *"Ochen Legkie Samolety"* (Russian: *Очень Легкие Самолеты (ОЛС)*) and means *"Very Light Airplanes"*.
- AP-VD stands for *"Vspomogatelnyi Dvigatel"* (Russian: *Вспомогательный Двигатель (ВД)*) and means *"Auxiliary Power Unit"*.

Appendix B

The full list and titles of the Federal Aviation Regulations (FAP) arranged according to responsible authorities (the authors' translation into English):

-- *Orders of the Government of the Russian Federation*

1. *Order 360 (1998).* About Federal Regulations of airspace use and Federal Aviation Regulations.
2. *Order 138 (2010).* Federal Regulations of Russian Federation airspace use.
3. *Order 530 (2008).* Federal Aviation Regulations of search and rescue in the Russian Federation.
4. *Order 303 (2000).* Rules of aircraft accident and incident investigation for Experimental Aircraft.
5. *Order 1329 (1999).* Rules of aircraft accident and incident investigation for Public Aircraft.
6. *Order 609 (1998).* Rules of aircraft accident and incident investigation for Civil Aircraft.

--- *Orders of the Ministry of Transport*

7. *Order 262 (2015).* Requirements to aerodromes intended for take-off, landing, taxiing, and parking of Civil aircraft.
8. *Order 251 (2015).* Procedure of state registration of Civil Aviation aerodromes and heliports.
9. *Order 250 (2015).* Procedure of landing ground owner's notification on commencement, suspension, or termination of landing ground operation for Civil aircraft flights, and registration in Civil Aviation authority.
10. *Order 216 (2015).* Requirements to juridical persons who implement aeronautical service for aircraft of Russian Federation airspace users.
11. *Order 297 (2014).* Radio communication support of aircraft flights and aircraft electrical communication in Civil Aviation.
12. *Order 60 (2014).* Meteorological information support for aircraft flight management.
13. *Order 32 (2014).* Requirements to Civil Aviation personnel license formatting.
14. *Order 362 (2012).* Implementation of radio communication in Russian Federation airspace.
15. *Order 6 (2012).* Russian Federation airspace use planning and management.
16. *Order 293 (2011).* Air transport management in the Russian Federation.
17. *Order 69 (2011).* Requirements to landing grounds located on land and water.
18. *Order 249 (2009).* Requirements to mandatory certification of natural and juridical persons performing aviation works. Certification procedure.
19. *Order 216 (2009).* Requirements to air traffic dispatchers and parachute instructors.
20. *Order 128 (2009).* Flight preparation and implementation in Civil Aviation.
21. *Order 147 (2008).* Requirements to air crew men, aircraft maintenance engineers, and flight support crew / flight dispatchers in Civil Aviation.
22. *Order 141 (2008).* Rules of dangerous cargo transportation by civil aircraft.
23. *Order 82 (2007).* General Regulations of passenger, baggage, cargo air transportation and requirements to passenger, consignor, consignee service.
24. *Order 142 (2005).* Aviation safety requirements to airports.
25. *Order 31 (2005).* Joint facilities of air traffic implementation.
26. *Order 1 (2005).* Flight checks of ground radio communication equipment, aircraft electrical communication, and light signaling equipment of Civil Aviation aerodromes.
27. *Order 202 (2003).* Technical means for Aviation work implementation. Requirements and certification procedures.
28. *Order 150 (2003).* Certification requirements to juridical persons engaged in airport works in passenger, baggage, cargo and mail service.
29. *Order 149 (2003).* Certification requirements to juridical persons engaged in airport works in electrical light support of flights.
30. *Order 147 (2003).* General Aviation operators. Requirements to General Aviation operators, procedures of registration and inspection of General Aviation operator service.
31. *Order 132 (2003).* Aircraft Example. Requirements and certification procedures.
32. *Order 118 (2003).* Rules of putting single aircraft examples into General Aviation service.
33. *Order 29 (2003).* Aviation safety requirements to General Aviation operators.
34. *Order 19 (2003).* Certification of ground aviation equipment.
35. *Order 126 (2002).* Certification requirements to organizations implementing quality check for aircraft fuel, oils, lubricants and special liquids.

--- *Orders of the Ministry of Defense*

36. *Order 265 (2009).* Medical support of Public Aviation flights.
37. *Order 60 (2009).* Unification of Public aviation personnel work conditions.
38. *Order 150 (2007).* Federal Aviation Regulations of Public Aviation Navigation.
39. *Order 275 (2004).* Federal Aviation Regulations of Public Aviation flight implementation.
40. *Order 460 (2002).* State registration of Public aircraft.
41. *Order 440 (2002).* State registration of Public Aviation aerodromes.

42. *Order 431 (2001).* Federal Aviation Regulations of ground service organization in Public Aviation.
43. *Order 420 (2001).* Organization of objective supervision in Public Aviation.
44. *Order 412 (2001).* Organization of training center in Public Aviation.
--------------------------------------- *Orders of the Russian Air Navigation Agency (abolished in 2009)*
45. *Order 119 (2007).* Placement of marking facilities on buildings, constructions, communication lines, electricity lines, radio equipment and other units installed for aircraft flight safety protection.
46. *Order 116 (2007).* Certification Regulations for joint facilities of air traffic implementation.
-- *Orders of the Russian Aerospace Agency (present name – State Corporation for Space Activities)*
47. *Order 165 (2003).* Organization of works of experimental aviation medical staff.
48. *Order 164 (2003).* Organization of flight medical center works for examination of experimental aviation.
49. *Order 80 (2003).* Flight medical examination of experimental aviation flight staff.
------- *Orders of the Federal Air Transport Service (present name – Federal Air Transport Agency)*
50. *Order 121 (2000).* Certification requirements to juridical persons who implement airport work in aerodrome support of aircraft flights.
51. *Order 89 (2000).* Certification requirements to organizations of aviation fuel supply for air transportation.
52. *Order 115 (1999).* Certification of juridical persons who implement medical examination of aviation staff.
53. *Order 10 (1999).* Mandatory certification of organizations engaged in supply of onboard food for passengers and air crew.
--- *Collective Order of Russian Ministries*
54. *Order 136/42/51 (2002) of the Ministry of Defense, Ministry of Transport, Russian Aerospace Agency.* Federal Aviation Regulations of flights in airspace of the Russian Federation.
-------------------------------------- *Orders commonly referred to as "FAP-XXX" with a serial number*
55. "FAP-16". *Order 397 (2007) of the Government.* About the improvement of state regulation of air carriers in air transportation.
56. "FAP-67". *Order 50 (2002) of the Ministry of Transport.* Medical examination of air crew, dispatchers, flight attendants, pilot students, and entrants to schools of Civil Aviation.
57. "FAP-119". *Order 246 (2015) of the Ministry of Transport.* Requirements to juridical persons, private entrepreneurs, who implement commercial air transportation.
58. "FAP-142". *Order 289 (2015) of the Ministry of Transport.* Requirements to educational institutions and organizations preparing aviation specialists of proper level.
59. "FAP-145". *Order 285 (2015) of the Ministry of Transport.* Requirements to juridical persons, private entrepreneurs, who implement maintenance of Civil Aviation aircraft.
60. "FAP-151". *Order 286 (2015) of the Ministry of Transport.* Requirements to operators of Civil Aviation aerodromes.
61. "FAP-173". *Order 270 (1998) of the Federal Aviation Service.* Radio communication support of aircraft flights and aircraft electrical communication. Certification requirements.

References

[1] Chernykh O and Bakiiev M 2016 Learning of Russian Aviation regulations by Chinese and international students *Proc. Asian Workshop on Aircraft Design Education* (Nanjing: Nanjing University of Aeronautics and Astronautics) pp 130–134
[2] Krasotkin A A 2007 *Sertifikatsiya Aviatsionnoy Tekhniki* [*Certification of Aircraft*] (Moscow: Moscow Aviation Institute) [in Russian]
[3] 27.03.1998 No.360 O Federal'nykh Pravilakh Ispol'zovaniya Vozdushnogo Prostranstva i Federal'nykh Aviatsionnykh Pravilakh [About Federal Regulations of Airspace Use and Federal Aviation Regulations] *Postanovlenie Pravitel'stva Rossiyskoy Federatsii* [*Order of the Government of the Russian Federation*] ed 18.02.2016 [in Russian]
[4] 2016 *Interstate Aviation Committee* http://www.mak-iac.org
[5] 2016 *Government of the Russian Federation* http://government.ru
[6] 2016 *Ministry of Transport of the Russian Federation* http://www.mintrans.ru

[7] 2016 *Ministry of Defense of the Russian Federation* http://mil.ru
[8] 2016 *Federal Air Transport Agency of the Russian Federation* http://www.favt.ru
[9] 2016 *Unified System of Aviation Standards* http://esas.niisu.ru
[10] 2013 *Aviatsionnye Pravila Chast' 21 Sertifikatsiya Aviatsionnoy Tekhniki, Organizatsiy Razrabotchikov i Izgotoviteley* [*Aviation Regulations Part 21 Certification Procedures for Aeronautical Products, Design Organizations and Manufacturers*] (Moscow: Interstate Aviation Committee) [in Russian]
[11] 25.01.2016 No.11 Ob Otmene Prikazov Ministerstva Transporta Rossiyskoy Federatsii [About Invalidation of Orders of the Ministry of Transport of the Russian Federation] *Prikaz Ministerstva Transporta Rossiyskoy Federatsii* [*Order of the Ministry of Transport of the Russian Federation*] [in Russian]
[12] 26.10.2015 No.316 Ob Otmene Prikaza Ministerstva Transporta Rossiyskoy Federatsii ot 06.11.2014 No.308 [About Invalidation of the Order of the Ministry of Transport of the Russian Federation from 06.11.2014 No.308] *Prikaz Ministerstva Transporta Rossiyskoy Federatsii* [*Order of the Ministry of Transport of the Russian Federation*] [in Russian]
[13] 12.09.2008 No.147 Ob Utverzhdenii Federal'nykh Aviatsionnykh Pravil "Trebovaniya k Chlenam Ekipazha Vozdushnykh Sudov, Spetsialistam po Tekhnicheskomu Obsluzhivaniyu Vozdushnykh Sudov I Sotrudnikam po Obespecheniyu Poletov (Poletnym Dispetcheram) Grazhdanskoy Aviatsii" [About Ratification of Federal Aviation Regulations "Requirements to Air Crew Men, Aircraft Maintenance Engineers, and Flight Support Crew (Flight Dispatchers) in Civil Aviation"] *Prikaz Ministerstva Transporta Rossiyskoy Federatsii* [*Order of the Ministry of Transport of the Russian Federation*] ed 16.09.2015 [in Russian]
[14] 2016 *Federal Aviation Administration* http://www.faa.gov
[15] 2016 *European Aviation Safety Agency* https://www.easa.europa.eu
[16] 2016 *International Civil Aviation Organization* http://www.icao.int
[17] 29.09.2015 No.289 Ob Utverzhdenii Federal'nykh Aviatsionnykh Pravil "Trebovaniya k Obrazovatel'nym Organizatsiyam i Organizatsiyam, Osushchestvlyayushchim Obuchenie Spetsialistov Sootvetstvuyushchego Urovnya Soglasno Perechnyam Spetsialistov Aviatsionnogo Personala" [About Ratification of Federal Aviation Regulations "Requirements to Educational Institutions and Organizations Preparing Aviation Specialists of Proper Level According to Aviation Specialist Lists"] *Prikaz Ministerstva Transporta Rossiyskoy Federatsii* [*Order of the Ministry of Transport of the Russian Federation*] [in Russian]
[18] 2014 *Vozdushnoe Zakonodatel'stvo, Sbornik* [*Aerial Legislation, Collection*] (Moscow: Aircraft Owners & Pilots Association) [in Russian]

IOP Conf. Series: Materials Science and Engineering **326** (2018) 012012 doi:10.1088/1757-899X/326/1/012012

A CFD Heterogeneous Parallel Solver Based on Collaborating CPU and GPU

Jianqi Lai[1,*], Zhengyu Tian[1], Hua Li[1] and Sha Pan[1]

[1]College of Aerospace Science and Engineering, National University of Defense Technology. Deya Road, Kaifu District, Changsha, Hunan, China

*laijianqi_kd@nudt.edu.cn

Abstract. Since Graphic Processing Unit (GPU) has a strong ability of floating-point computation and memory bandwidth for data parallelism, it has been widely used in the areas of common computing such as molecular dynamics (MD), computational fluid dynamics (CFD) and so on. The emergence of compute unified device architecture (CUDA), which reduces the complexity of compiling program, brings the great opportunities to CFD. There are three different modes for parallel solution of NS equations: parallel solver based on CPU, parallel solver based on GPU and heterogeneous parallel solver based on collaborating CPU and GPU. As we can see, GPUs are relatively rich in compute capacity but poor in memory capacity and the CPUs do the opposite. We need to make full use of the GPUs and CPUs, so a CFD heterogeneous parallel solver based on collaborating CPU and GPU has been established. Three cases are presented to analyse the solver's computational accuracy and heterogeneous parallel efficiency. The numerical results agree well with experiment results, which demonstrate that the heterogeneous parallel solver has high computational precision. The speedup on a single GPU is more than 40 for laminar flow, it decreases for turbulent flow, but it still can reach more than 20. What's more, the speedup increases as the grid size becomes larger.

1. Introduction

With the rapid development of computer technology, computational fluid dynamics (CFD) plays an important role in aerodynamic performance analysis, optimal design and complex flow mechanism study. At the same time, the requirements of engineering calculation in CFD are also increasing. As for complete aircraft configuration, the large scale simulations on grids consisting of dozens of millions of elements, the huge amount of calculation could only be completed by parallel computing in computer cluster or supercomputer. Graphic Processing Unit (GPU) has a strong ability of floating-point computation and memory bandwidth for data parallelism. It has been widely used in the areas of common computing [1, 2] such as molecular dynamics (MD), computational fluid dynamics (CFD) and so on. The emergence of compute unified device architecture (CUDA) brings great opportunities to CFD, which could promote the development of CFD.

Traditionally, GPU was limited to the area of graphics rendering. Compute unified device architecture (CUDA) is the software and hardware architecture for GPU produced by NVIDIA in 2007, which reduce the complexity of compiling program [3, 4]. The emergence of CUDA as a programming model marked the beginning of widespread extension of GPUs to scientific computing. In the field of CFD, a series of important results have been achieved in the implementation of large-scale parallel computation on GPU cluster [5-7]. Some scholars started to solve Euler equations and Navier-Stokes equations on GPU cluster, further research on Reynolds-averaged equations (RANS), Large Eddy Simulation (LES) and Direct Numerical Simulation (DNS) were also widely carried out, and the computational efficiency had been greatly improved.

In this paper, a heterogeneous parallel solver based on collaborating CPU and GPU is established in parallel cluster. The reminder of this paper is organized as follows. Section 2 introduces the governing equations and numerical schemes. Section 3 discusses the GPU hardware and the detailed GPU implementation. Section 4 presents several numerical results to verify the solver's computational precision in laminar and turbulent flow. Section 5 gives the performance analysis of the solver. Finally in section 6, we introduce some final conclusions and directions for future work.

2. Governing equations

Without regard to volume sources due to body forces and volumetric heating, Navier-Stokes equations for compressible flow can be written as follows [8]:

$$\frac{\partial}{\partial t}\int_{\Omega}\vec{W}d\Omega+\oint_{\partial\Omega}(\vec{F_c}-\vec{F_v})dS=0 \tag{1}$$

$$\vec{W}=\begin{bmatrix}\rho\\ \rho u\\ \rho v\\ \rho w\\ \rho E\end{bmatrix}\qquad \vec{F_c}=\begin{bmatrix}\rho V\\ \rho uV+n_x p\\ \rho vV+n_y p\\ \rho wV+n_z p\\ \rho HV\end{bmatrix}\qquad \vec{F_v}=\begin{bmatrix}0\\ n_x\tau_{xx}+n_y\tau_{xy}+n_z\tau_{xz}\\ n_x\tau_{yx}+n_y\tau_{yy}+n_z\tau_{yz}\\ n_x\tau_{zx}+n_y\tau_{zy}+n_z\tau_{zz}\\ n_x\Theta_x+n_y\Theta_y+n_z\Theta_z\end{bmatrix}$$

Where $\vec{W}$ is the vector of conservative variables, $\vec{F_c}$ is the vector of convective fluxes, $\vec{F_v}$ is the vector of viscous fluxes, and Ω is a control volume.

The spatial discretisation of these equations on structured grids is using a cell-centered finite volume scheme as shown in figure 1. Compared to the unstructured grids, structured grids can easily achieve better performance due to its body fitted property.

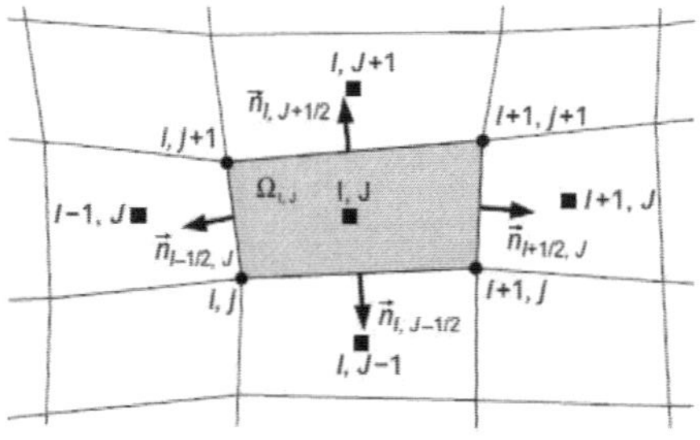

Figure 1. Control volume of a cell-centered scheme.

For a control volume $\Omega_{I,J,K}$, the discrete form of governing equations separately for temporal and spatial can be written as:

$$\Omega_{I,J,K}\frac{d\vec{W}}{dt}+\sum_{m=1}^{N_F}(\vec{F_c}-\vec{F_v})\Delta S_m=0 \tag{2}$$

The convective fluxes are computed by AUSM+UP scheme, and the viscous fluxes are computed by central scheme, the Van Leer's limiter is used to prevent the generation of oscillations and spurious solutions in regions of large gradient. Here, explicit multi-stage Runge-Kutta scheme for the temporal discretization of governing equations are discussed. What's more, the $k-\omega$ SST two-equation model is used to solve the turbulence source term.

3. Heterogeneous parallel solver collaborating CPU and GPU

3.1. GPU hardware

The GPU used in this paper is NVIDIA GTX 1070 which has 8GB device memory. GTX 1070 contains 15 streaming multiprocessors (SMs) and each SM has 192 CUDA cores. The GTX 1070's compute capability is 6.1, which could meet the demand of large-scale scientific computation. The large memory size and strong compute capability makes GTX 1070 attractive for the solution of incompressible or compressible Navier-Stokes equations. Table 1 shows the main performance parameters of GPU.

Table 1. The main performance parameters of GTX 1070.

GPU Parameters	GTX 1070
Device Memory	8GB
Compute Capability	6.1
Single-Precision Float	6080 Gflop/s
Double-Precision Float	194 Gflop/s
Bandwidth	256.3GB/s

3.2. Parallel Programming Model of CUDA

The entire CUDA program contains both host (CPU) and device (GPU) as shown in figure 2, the part of host program executes on the CPU, while the other which calls kernel executes on the GPU. A kernel is a function running on the GPU, corresponds to a thread grid. A thread grid is executed by threads, which are grouped into batches of up to 1024 called thread block, Threads within the same thread block are guaranteed to run on the same SM at the same time [9, 10]. In CUDA, a thread grid can be divided into a number of blocks, while the number of blocks and the number of threads per block will have a great influence on the computational performance of GPU, we should do some tests to determine appropriate number of blocks in a thread grid and threads per block.

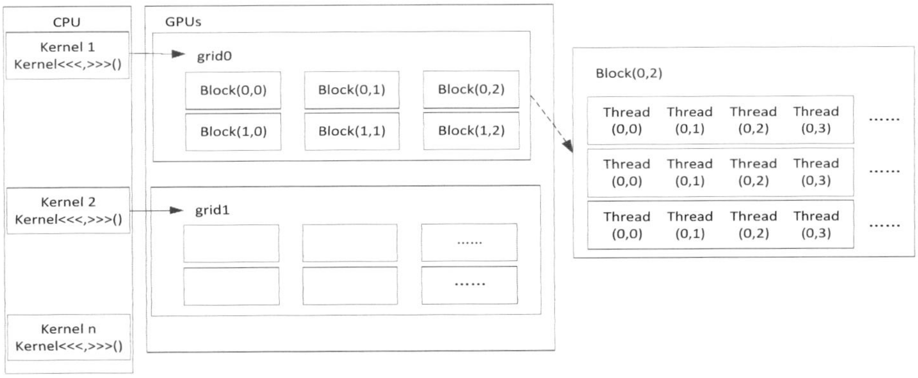

Figure 2. Parallel programming model of CUDA.

3.3. Heterogeneous parallel solver

Graphic Processing Unit (GPU) has a strong ability of floating-point computation and memory bandwidth for data parallelism. As we can see, GPU is relatively rich in compute capacity but poor in memory capacity and the CPU do the opposite. So the CPU is suitable for dealing with logical affairs such as selection, judgement and so on, while the GPU is good at handling with single instructions and other highly linearized tasks. In general, there are three different modes for parallel solution of NS equations: parallel solver based on CPU, parallel solver based on GPU and heterogeneous parallel solver based on collaborating CPU and GPU [11, 12].

The basic idea of multi-CPU parallel mode is shown in figure 3. It is easy to be understood. Firstly, the computational domain should be divided into several subregions of which the grid sizes are equal, and then each CPU core load the computation of a subregion. However, the conventional parallel mode is hard to meet the requirements of large-scale computing in industry. Sometimes the task can't be implemented because of the extensive calculation. As for GPU-only parallel mode, we can't make the best of its strong computation ability because of the weak memory capacity of GPU. In order to make full use of the GPUs and CPUs, a CFD heterogeneous parallel solver based on collaborating CPU and GPU has been established.

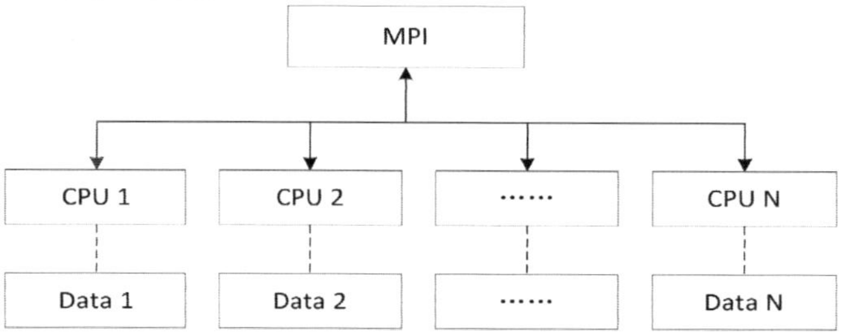

Figure 3. Multi-CPU parallel mode.

Parallel mode based on collaborating CPU and GPU as shown in figure 4. According to the characteristics of NS equations in CFD, the computational task could be divided into branch instruction, single instruction and data transfer. The branch instruction task such as the treatment of boundary conditions and data transfer are implemented by CPU, while the single instruction task such as the calculation of fluxes in the inner face are implemented by GPU. It is a good choice for researchers to combine the characteristics of computing task with device.

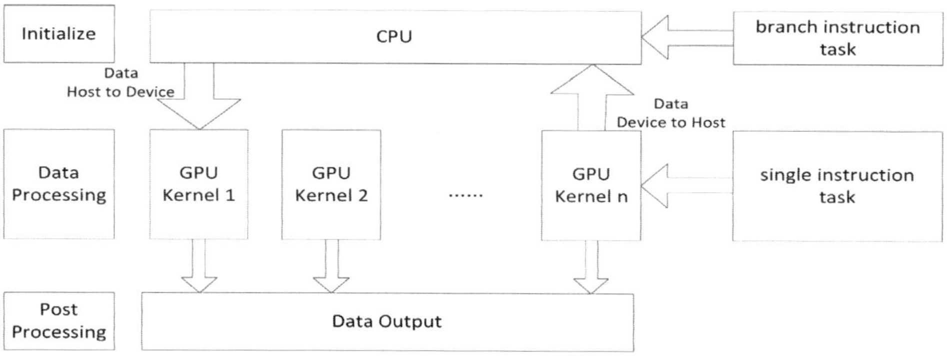

Figure 4. Parallel mode based on collaborating CPU and GPU.

In this paper, we use CUDA of version 7.5, Intel icc11.1 for C code. A single node platform contains a GPU of NVIDIA GTX 1070 and a CPU of Intel Xeon 2670 with eight cores. It's important to improve single node performance before realizing multi-node version. A node contains several GPUs or multi-node version will be discussed in future work.

4. Numerical results

In this section, three cases are presented to test the computational accuracy of heterogeneous parallel solver [13, 14]. For these three cases, the pressure distribution of symmetry plane is chosen to analyse the solver's accuracy.

4.1. Double Ellipsoid

This case demonstrates the solver's ability to simulate laminar, hypersonic flow over the double ellipsoid. A large number of wind tunnel experiments for the model were implemented with different Mach numbers and angle of attacks in Ref. 13, and a wind tunnel experiment result is used as a comparison.

Free flow conditions of this case are given in table 2.

Table 2. Free flow conditions for double ellipsoid.

Ma	Re/m^{-1}	P_0 / KPa	T_0 / K	Angle of Attack(deg)
8.02	1.98×10^7	850	720	0

Pressure distribution of the upper and lower surfaces is shown in figure 5 and figure 6. The numerical results agree well with experiment results, which demonstrate that the heterogeneous parallel solver has high computational precision.

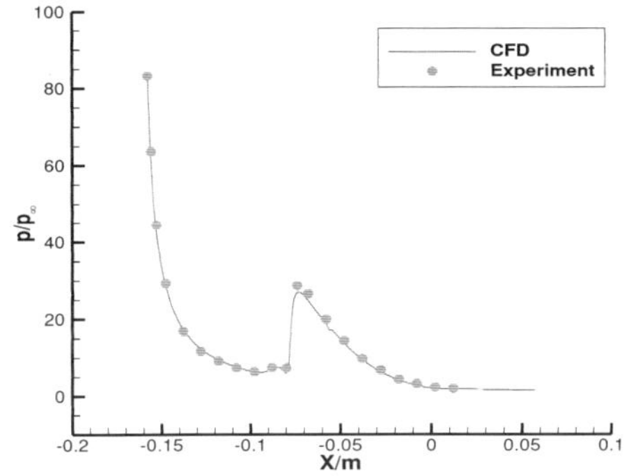

Figure 5. Pressure distribution of upper surface.

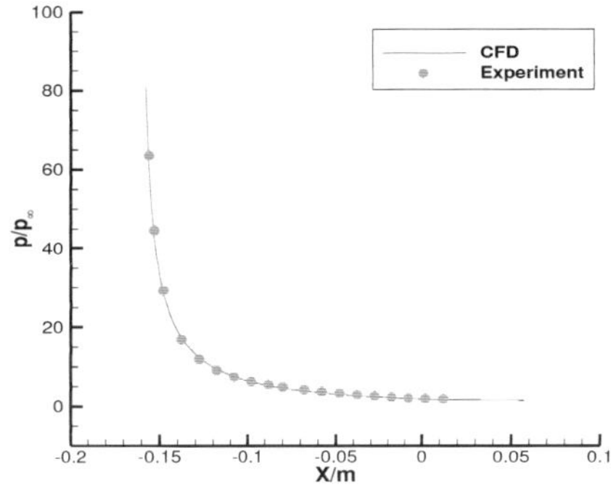

Figure 6. Pressure distribution of lower surface.

4.2. Aerospace plane

This case demonstrates the solver's ability to simulate laminar, hypersonic flow over the aerospace plane which is significantly important to future space battle and transportation. The same as the first case, a wind tunnel experiment result from Ref. 13 is used as a comparison.

Free flow conditions of this case are given in table 3.

Table 3. Free flow conditions for aerospace plane.

Ma	Re/m^{-1}	P_0 / KPa	T_0 / K	Angle of Attack(deg)
8.02	1.34×10^7	600	740	0

Pressure distribution of the upper and lower surfaces a shown in figure 7 and figure 8. The numerical results agree well with experiment results, showing the heterogeneous parallel solver has strong ability to deal with complex geometries.

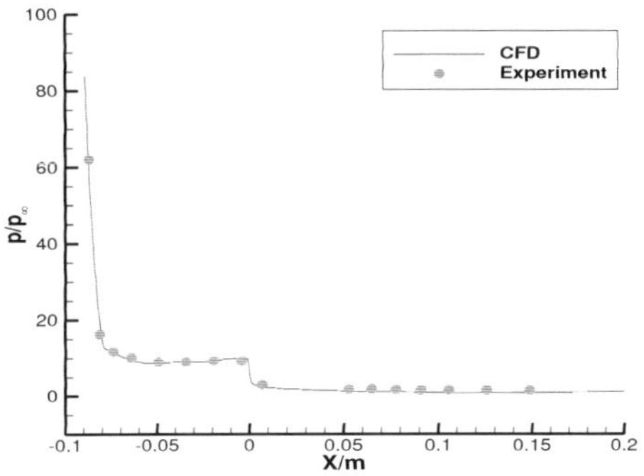

Figure 7. Pressure distribution of upper surface.

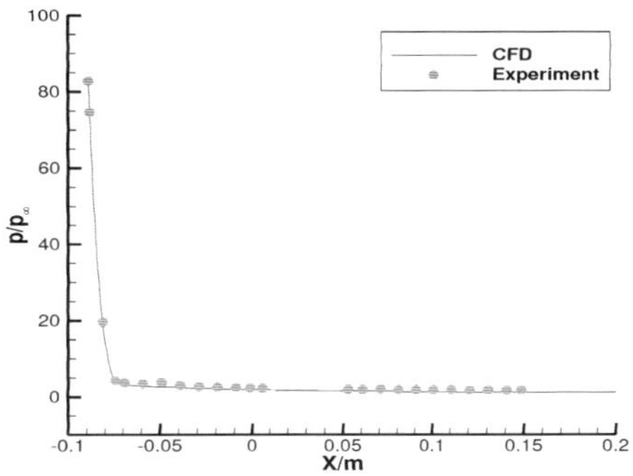

Figure 8. Pressure distribution of lower surface.

4.3. Supersonic inlet

This case demonstrates the solver's ability to simulate turbulent, supersonic flow over the supersonic inlet. The supersonic inlet is of great importance in the airbreathing propulsion system. The flow field

of supersonic inlet is very complex that contains boundary layer, expansion wave, contour induced shocks, reflected and boundary layer induced shocks.
Free flow conditions of this case are given in table 4.

Table 4. Free flow conditions for supersonic inlet.

Ma	$Re\,/\,\mathrm{m}^{-1}$	P_0 / KPa	T_0 / K	Angle of Attack(deg)
2.41	5.07×10^{7}	540	305	-10

In this case, computing mesh was generated by the software of Gridgen, and structured grid was utilized here. Computational domain and grid detail for supersonic inlet is shown in figure 9.

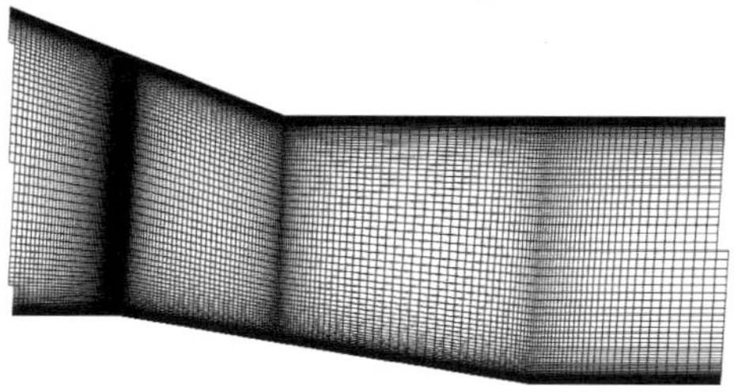

Figure 9. Computational domain and grid detail for supersonic inlet

Figure 10 and figure 11 shows the comparison of a schlieren picture from Ref. 14 with corresponding Mach number contour lines of the computation for supersonic inlet, and it reveals a good agreement.

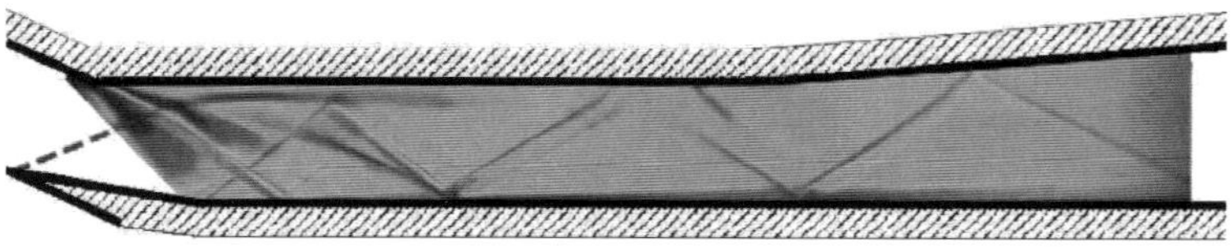

Figure 10. A schlieren picture for supersonic inlet

Figure 11. Mach number contour lines of the computation for supersonic inlet

Pressure distribution of the upper and lower surfaces is shown in figure 12 and figure 13. A quantitative comparison between numerical and experiment results demonstrate that the solver performs well in solving turbulence problems.

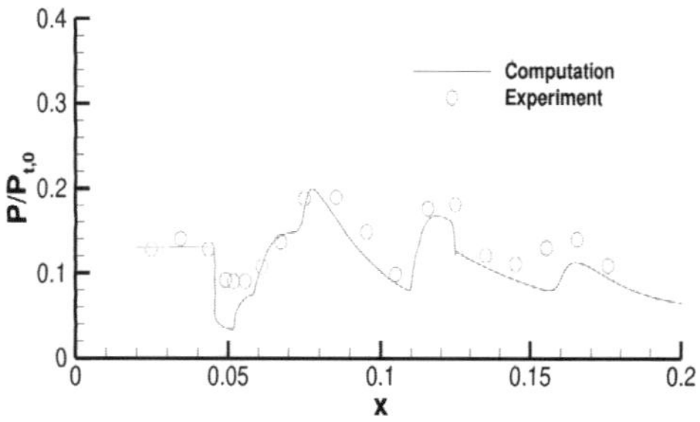

Figure 12. Pressure distribution of upper surface.

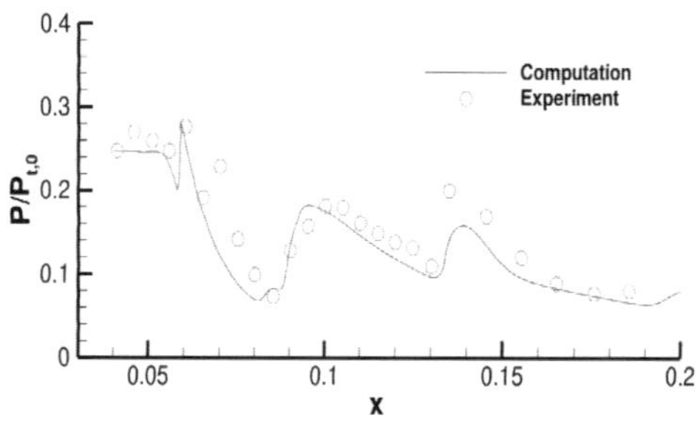

Figure 13. Pressure distribution of lower surface.

5. Performance analysis

In order to apply the solver to large-scale parallel computation, we must study the heterogeneous parallel efficiency of the solver. Speedup is an important performance parameter to evaluate parallel efficiency. It is defined as the CPU operation time divided by the heterogeneous parallel operation time while implemented in the same conditions, here the CPU operation time is used as a reference to compute heterogeneous parallel efficiency. To collect the operation time, we simulate 10 time steps, with each having 100 sub_iterations, and the average time per iteration step is chosen as the operation time.

The speedup varies with gridsize for double ellipsoid and aerospace plane can be shown in figure 14 and figure 15, and figure 16 shows that the speedup varies with gridsize for supersonic inlet. The solver achieves a speedup of more than 40 in laminar flow, it decreases when handling with turbulence flow, but it still can reach more than 20. The results demonstrate that the heterogeneous parallel solver has high computational efficiency. What's more, for the three cases, the speedup increases as the grid size becomes larger. This is due to the proportion of single instruction task increases compares with branch instruction task. As we all know, GPUs specialize in dealing with single instruction task, while the CPUs are good at dealing with branch instruction task. So the heterogeneous parallel solver takes full advantages of the characteristics of hardware, and it provides a new direction for scientific computing.

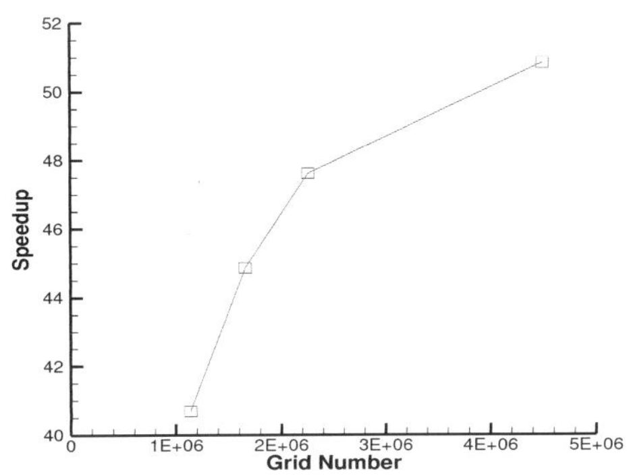

Figure 14. Speedup for double ellipsoid

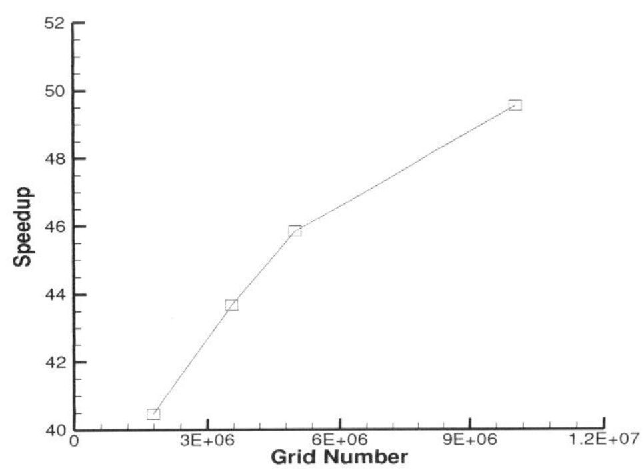

Figure 15. Speedup for aerospace plane

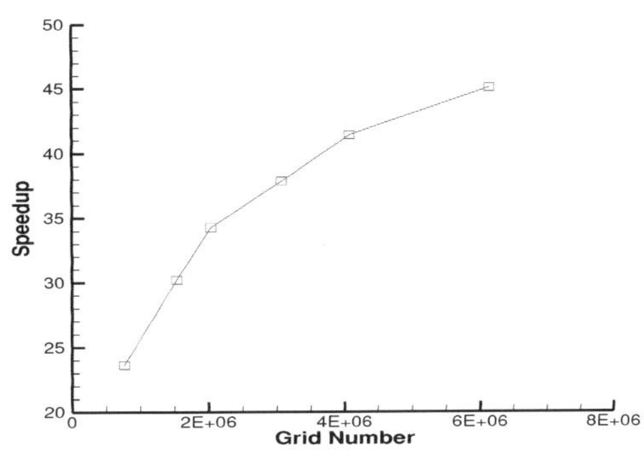

Figure 16. Speedup for hypersonic inlet.

6. Conclusion

Heterogeneous parallel solver based on collaborating CPU and GPU was established on parallel computing cluster in this paper. The numerical results agree well with experiment results, which demonstrate that the solver has high computational precision in laminar and turbulent flow. The speedup on a single GPU is more than 40 for laminar flow, it decreases for turbulent flow, but it still can reach more than 20. For the three cases, the speedup increases as the grid size becomes larger because the proportion of single instruction task increases compares with branch instruction task. A multi-GPU heterogeneous parallel solver will be discussed in future work.

References

[1] Esfahanian V, Baghapour B, Torabzadeh M, and Chizari H 2014 An efficient GPU implementation of cyclic reduction solver for high-order compressible viscous flow simulations *Computers & Fluids* **92** 160-71
[2] Liu X, Zhong Z and Xu K 2016 A hybrid solution method for CFD applications on GPU-accelerated hybrid HPC platforms *Future Generation Computer Systems* **56** 759-65
[3] NVIDIA Corporation 2013 *CUDA C Programming Guide*
[4] Dana A J, Julien C T, Thibault and Inanc S 2010 An MPI-CUDA Implementation for Massively Parallel Incompressible Flow Computations on Multi-GPU Clusters *48th AIAA Aerospace Sciences Meeting Including the New Horizons Forum and Aerospace Exposition* p 522
[5] Brandvik T and Pullan G 2008 Acceleration of a 3D Euler solver using commodity graphics hardware *46th AIAA Aerospace Sciences Meeting and Exhibit* p 607
[6] Watkins J E, Romero J and Jameson A 2016 Multi-GPU, implicit time stepping for high-order methods on unstructured grids *46th AIAA Fluid Dynamics Conf.* p 3965
[7] Corrigan A, Camelli F F, Löhner R and Wallin J 2011 Running unstructured grid-based CFD solvers on modern graphics hardware *International Journal for Numerical Methods in Fluids* **66** 221-29
[8] Blazek J 2005 *Computational Fluid Dynamics: Principles and Applications (2nd Edition)* (Amsterdam: Elsevier)
[9] Castonguay P, Williams D, Vincent P, Lopez M and Jameson A 2011 On the development of a high-order, multi-GPU enabled, compressible viscous flow solver for mixed unstructured grids *20th AIAA Computational Fluid Dynamics Conf.* p 3229
[10] Kampolis I, Trompoukis X, Asouti V and Giannakoglou K C 2010 CFD-based analysis and two-level aerodynamic optimization on graphics units *Computer Methods in Applied Mechanics and Engineering* **199** 712-22
[11] Xu T and Chen L 2016 GPU implementation of a viscous flow solver on unstructured grids 6th *Int. Symp. on Physics of Fluids* vol 42
[12] Xu C et al. 2014 Collaborating CPU and GPU for large-scale high-order CFD simulations with complex grids on the TianHe-1A supercomputer *Journal of Computational Physics* **278** 275-97
[13] Li S 2007 *Hypersonic Flow Characteristics of Typical Shape* (Beijing: National Defense Industry Press)
[14] Reinartz B U, Herrmann C D and Ballmann J 2003 Aerodynamic performance analysis of a hypersonic inlet isolator using computation and experiment *Journal of Propulsion & Power* **19** 868-75

IOP Conf. Series: Materials Science and Engineering **326** (2018) 012013 doi:10.1088/1757-899X/326/1/012013

Experimental Investigation of Two-Phase Oil (D130)-Water Flow in 4" Pipe for Different Inclination Angles

S. M. Shaahid[1,*], Mehaboob Basha[1] and Luai M. Al-Hems[1]

[1]Centre for Engineering Research- Research Institute, King Fahd University of Petroleum and Minerals, Dhahran, Saudi Arabia

*mshaahid@kfupm.edu.sa

Abstract. Oil and water are often produced and transported together in pipelines that have various degrees of inclination from the horizontal. The flow of two immiscible liquids oil and water in pipes has been a research topic since several decades. In oil and chemical industries, knowledge of the frictional pressure loss in oil-water flows in pipes is necessary to specify the size of the pump required to pump the emulsions. An experimental investigation has been carried out for measurement of pressure drop of oil (D130)-water two-phase flows in 4 inch diameter inclined stainless steel pipe at different flow conditions. Experiments were conducted for different inclination angles including; 0°, 15°, 30° (for water cuts "WC" 0 - 100%). The flow rates at the inlet were varied from 4000 to 8000 barrels-per-day (BPD). For a given flow rate the frictional pressure drop has been found to increase (for all angles) from WC = 0 - 60 %, and thereafter friction pressure drop decreases, this could be due phase inversion. For a given WC 40%, the frictional pressure drop has been found to increase with angle and flow rate. It has been noticed that inclination angle has appreciable effect on frictional pressure drop.

1. Introduction

The multiphase flow is a complex phenomenon involving simultaneous flow of two or more physically immiscible fluids (such as: oil and water) in pipelines. Also, the widespread occurrence of multiphase flows in pipes is the driving force for extensive research in this area (a number of upstream practical applications in the petroleum industry involve oil–water two-phase flow phenomena).

The physical understanding of two-phase flow characteristics in pipes is of importance since appreciable savings in the pumping power required for oil transportation (water-lubricated transportation of crude oil) can be attained when water flows in the pipeline together with the oil, especially when the highly viscous phase is surrounded by a water annulus, giving place to the core annular flow configuration. More importantly, fluids with different properties exhibit different flow regimes in different pipe configurations under different operating conditions. Considerable literature exists on the two-phase flow of oil and water.

Jing et al. [1] investigated phase inversion and frictional pressure gradients during simultaneous vertical flow of oil and water two-phase through upward and downward pipes. They concluded that the frictional pressure gradient reaches to its lower value at the phase inversion point. Pedram et al. [2] investigated flow patterns of two-phase oil–water flow in an inclined pipe. They inferred that non-stratified flows such as bubbly and slug flows are dominant flow patterns in the upward flows and stratified flows are dominant flow patterns in the downward flows. Descamps et al. [3] performed investigation of three phase flow in vertical pipes. Attention was paid to phase inversion phenomenon.

They noticed that the dispersed water phase has significant impact on the bubble size. Grassi et al. [4] conducted experiments of two phase liquid-liquid (high viscosity ratio) flows in horizontal and slightly inclined pipes. The results were validated against theoretical models. Du et al. [5] conducted experimental investigation of vertical upward oil-water two phase flows in a 20 mm diameter pipe. They presented flow pattern map of oil water for different superficial velocities. Domenico et al. [6] conducted experimental study on oil/water flow in horizontal and slightly inclined small pipe plexi-glass tubes (with 21 mm ID, 9m long). They focused on core-annular flow pattern boundary, pressure drops, and oil hold-up measurements.

Karolina et al. [7] investigated phase inversion and its effect on pressure gradient of immiscible (water and oil) liquids for two pipe materials and two pipe sizes for a range of mixture velocities. Phase inversion was observed in all cases preceded by a large increase in pressure gradient.

Dongian et al. [8] studied local flow characteristics of oil–water dispersed flow in a vertical upward pipe. The water flow rates varied from 0.12 m/s to 0.89 m/s, while the oil flow rates ranged from 0.024 m/s to 0.198 m/s. The typical radial profiles of interfacial area concentration, oil phase fraction, interfacial velocity, and oil pressure drops were presented. Lum et al. [9] investigated the effect of upward and downward pipe inclinations on the flow patterns, hold up and pressure gradient during two-liquid (water and oil) phase flows for mixture velocities between 0.7 and 2.5 m/s and phase fractions between 10% and 90%. The oil to water velocity ratio was higher for the upward than for the downward flows but in the majority of cases and for all inclinations oil was flowing faster than water.

In the wake of the above literature review, currently no studies are available on frictional pressure drop (FPD) measurements of oil (D130)-water two-phase flow in inclined 4 inch diameter stainless steel pipe at different inclinations (at different flow conditions). This is the motivation for the present experimental study and it places emphasis on the effect of flow rates, water cuts, and inclination angles on pressure drop measurements of oil (D130)-water two-phase flow. In this work, attention has been focused on FPD measurements of oil (D130)-water two-phase flow in an inclined 4 inch diameter stainless steel pipe at different inclinations (at different flow conditions). Experiments were carried out for different inclination angles including; 0°, 15°, 30° (for water cuts 0%, 20%, 40% 60% 100%). The oil-water flow rates at the inlet were varied from 4000 to 8000 BPD in steps of 2000. The above information is helpful in effectively handling the frictional pressure loss issues.

2. Experimental setup

The Oil-water two phase experiments were conducted in the flow loop of the multiphase flow laboratory of King Fahd University of Petroleum and Mineral, Dhahran, Saudi Arabia. Details of the loop components and instruments are given in table 1.

The schematic diagram of the flow loop is shown in figure 1. Experimental set-up includes: four centrifugal variable speed pumps [2 pumps for water (WP) and 2 pumps for oil, (OP)], 4 inch stainless loop, a horizontal separator tank (WOST), which acts as storage tank, two level indicators for oil and water each. The loop is constructed on swinging platform (inclination can be varied from 0° - 30°). The flexible connection (FC) helps in positioning loop at any given angle. The loop is instrumented with a turbine type oil flow meter (OFM), a turbine type water flow meters (WFM), line pressure transmitter (LPT), two flow differential pressure transmitters (DP1 and DP2).

Table 1. Details of equipment of the flow loop.

Items	Manufacturer	Model	Capacity/Range	Accuracy/Error
Four pumps (two water, two oil)	NEWAR FLOW SERVE	50-32CPX200	35 m^3/hr	-
Two turbine flow meters	Omega	EF10	±10 m/s	±1.0 %
Line pressure gauge	ROSEMOUNT	AOB-20	0-7 bar	±0.25%
DPT1	ROSEMOUNT	300S2EAE5M9	0-70 "of H_2O	±0.1%
DPT2	ROSEMOUNT	300S2EAE5M9	0-12 "of H_2O	±0.1%

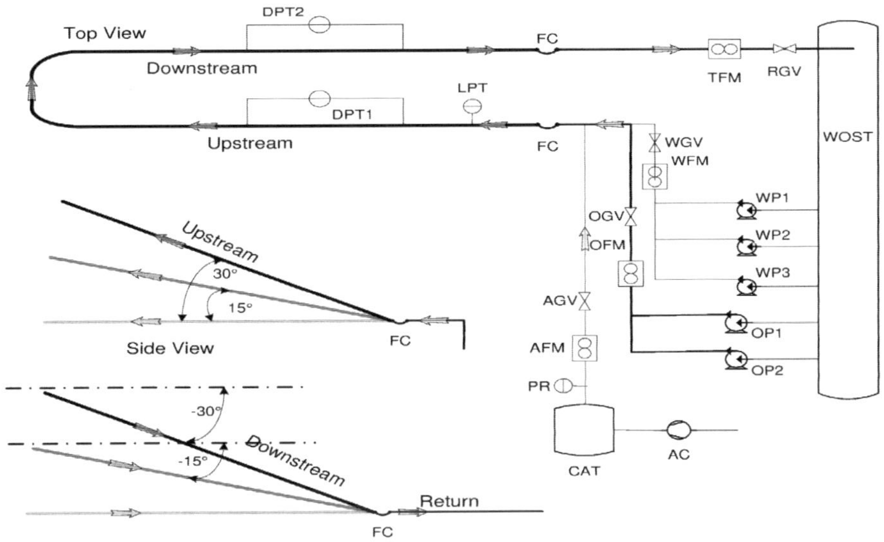

Figure 1. Schematic layout of the oil-water multiphase flow loop

3. Experimental procedure

In-order to validate the pressure drop measurements against available empirical models, experiments were performed for water-only and oil-only single phase (in 4inch pipe).

To achieve the above, water was pumped in the loop using centrifugal pumps. Required volume flow rate was attained by varying speed of pumps through variable speed drives and also by regulating oil globe valve (OGV) and water globe valve (WGV) of oil and water flow streams respectively. Turbine flow meters installed on the discharge line of the pumps were used for measuring the flow rates. Return gate valve (RGV, figure1) of the loop is throttled to set the required outlet pressure (e.g. 1 bar or 2 bars).

For a given flow rate, experiments were conducted and pressure drop measurements were made at different locations of the loop as shown figure 1. Once the steady state flow condition is achieved, differential pressure drops are recorded across 3m (DPT1). CR 1000 data logger was used to record experimental data. Similar procedure was followed for oil-only flow experiments.

Pressure drop data was used to calculate friction factor using Eq. (1) and compared with Eq. (2) and Eq. (3).

$$f = \frac{\Delta P}{L}\frac{2D}{\rho v^2} \quad (1)$$

ΔP	Pressure drop (Pa).	v	Average velocity of the fluid (m/s)
L	Distance between the two pressure taps (m)	ε	Pipe roughness (m)
		Re	Reynolds number
D	Hydraulic diameter of the pipe (m)		
ρ	Fluid density (Kg/m^3)		

$$f = 0.3164\, Re^{-1/4} \quad (2)$$

$$\frac{1}{\sqrt{f}} = -2\log\left[\frac{\varepsilon/D}{3.7} - \frac{5.02}{Re}\log\left[\left(\frac{\varepsilon/D}{3.7}\right) + \frac{13}{Re}\right]\right] \quad (3)$$

The turbulent friction factor can also be determined using other correlations, such as the Zigrang & Sylvester 1985 correlation defined in equation (3) above.

Then experimentally obtained friction factor (Eq. 1) was compared with the friction factors calculated by using Blasius correlation and Zigrang & Sylvester correlations as shown in the figure 2. The results showed a close agreement particularly with the Blasius friction factor (Eq. 2).

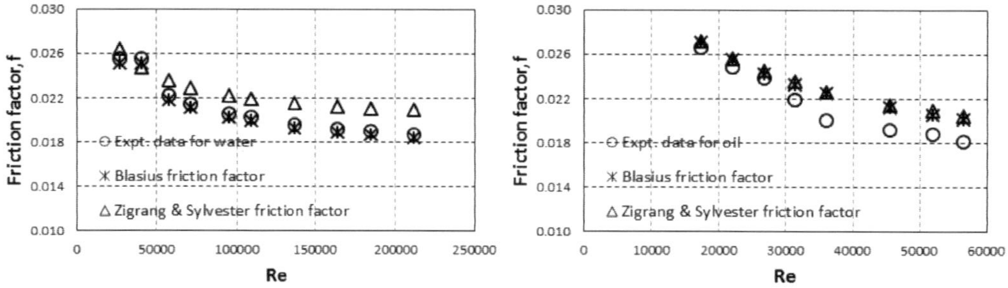

Figure 2. Friction factor comparisons with Blasius and Zigrang & Sylvester correlations for oil-water flow

Figure 2 shows the friction factor of single phase water and oil against Re. It can be noticed that the friction factor decreases with increase in velocity. The experimental data is found to be in good agreement with established theoretical relation. The above experiments were for single phase oil only and water only. However, for a given oil-water multiphase flow, speeds of the oil and water pumps was varied to achieve required flow rate and water cut. Once the required water cut and flow rates are reached, pressure drop [across 3m (DPT1)] measurements were made. Similar procedure was followed for other angles including; 15°, 30° and for different water cut ratio 0 to 100%. The oil-water flow rates at the inlet were varied from 4000 to 8000 BPD.

4. Results and discussions

Oil-water two phase flow experiments were carried out for different inclination angles including; 0°, 15°, 30° and for different water cut ratios (0%, 20%, 40%, 60% and 100%). The oil-water flow rates at the inlet were varied from 4000 to 8000 BPD.

4.1. Effect of water-cut on oil-water frictional pressure drop (FPD) for different flow rates

For a given angle θ = 15° case, the effect of water cut for different flow rates on pressure drop is shown in Figure 3a. As it can be seen from figure 3a, for a given flow rate the pressure drop increases from WC = 0 to WC 60 %. Further increase in WC, shows decrease in FPD has been found to decrease. This could be due to phase inversion or change in flow pattern regime. Also, it can be seen from figure 3a, for any given WC, the FPD increases with increase in flow rate.

For θ = 30° case, the effect of flow rate on FPD for different water cuts is shown in figure 3b. As it can be seen from figure 3b, FPD increases with flow rate and WC. However, FPD is relatively higher for WC 20 - 60% as compared θ = 15° case. This could be due to increase in inclination angle.

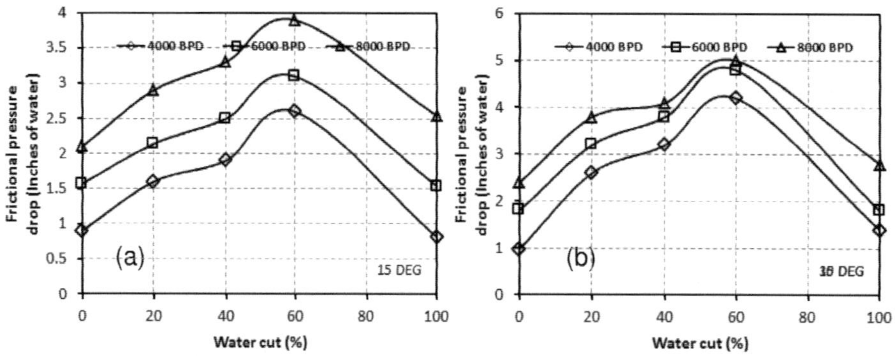

Figure 3. Effect of water cut on pressure drop for different flow rates. (a)15° case, (b) 30° case

4.2. Effect of flow rate on oil-water pressure drop for different water-cuts

For a given angle θ = 15°, the effect of flow rate on frictional pressure drop (FPD) for different WC is shown in figure 4a. As it can be seen from figure 4a, pressure drop increases with flow rate and WC. The FPD drop has been found to increase linearly with respect flow rate. For a given flow rate of 6000 BPD, for increase in water cut from WC 0 to 20, percentage increase in FPD is about 36%. For angle θ = 30° case, the effect of flow rate on FPD for different water cuts is shown in figure 4b. Again, as it can be seen from figure 4b, FPD increases with flow rate and WC. However, pressure drops are relatively higher as compared θ = 15° case. This could be due to increase in inclination angle. For a given flow rate of 6000 BPD, for increase in water cut from WC 0 to 20, percentage increase in frictional pressure drop is about 77%. From the these figures, it can be concluded that the effect of inclination on pressure drop behaviour is appreciable.

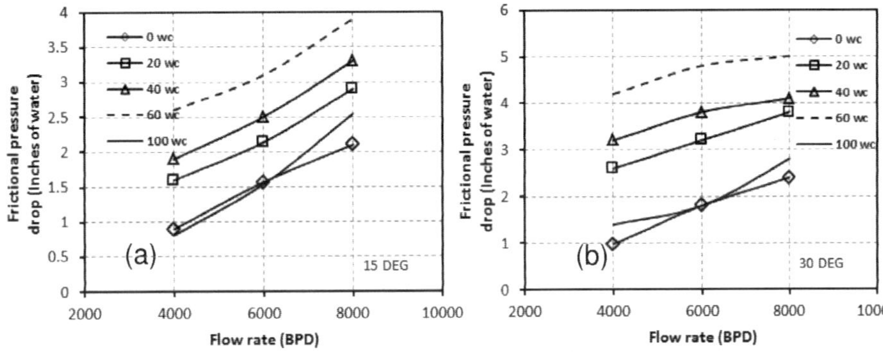

Figure 4. Effect of flow rate on pressure drop for different water cuts (a) 15° case, (b) 30° case

4.3. Effect of inclination on oil-water pressure drop for different flow rates for given water cut
For the sake of brevity, the angle effect on pressure drop measurements for different flow rates, only water cut (WC) = 40% has been presented.

For a given water cut (WC = 40%), the effect of inclination for different flow rates on pressure drop is shown in figure 5. As mentioned earlier, in general for all angles, pressure drop increases with flow rate and water cut. The effect of angle has found to be appreciable. For a given flow rate 8000 BPD, WC = 40%, increase in angle from 0 to 15°, percentage increase in frictional pressure drop is about 50%. However, for further increase in angle from 15° to 30°, percentage increase in frictional pressure drop is about 24%.

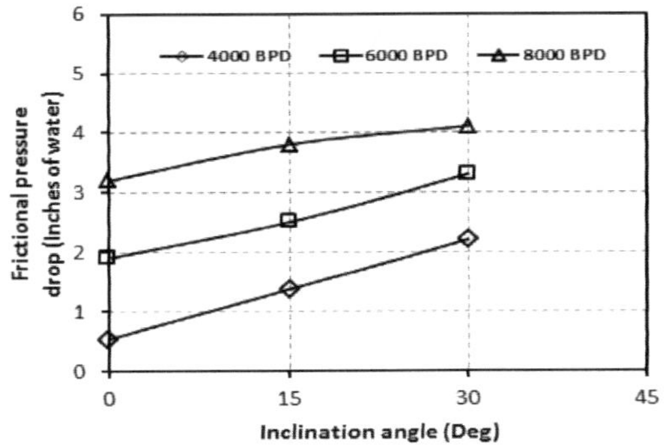

Figure 5. Effect of angle on pressure drop for different flow rates (for a given water cut, WC = 40%)

The total pressure head (TPH) is sum of frictional and gravitational pressure head (GPH). The present work has focused on frictional pressure head (FPH). However, GPH is a constant term which may be added to FPH to obtain the TPH. For a given angle, θ = 15°, GPH (ρgh = ρ g sinθ h, ρ is mixture density, h is distance between pressure tapping points, g is gravity) is 23.13 inches of water.

5. Conclusions

The present work has focused attention on the pressure drop measurements of oil (D130)-water two-phase flow in a horizontal and inclined 4 inch diameter stainless steel pipe at different flow conditions. Experiments were performed for different inclination angles including; 0°, 15°, 30° and for different water cut ratios (0%, - 100%). The flow rates at the inlet were varied from 4000 to 8000 BPD. In order to validate the experimental work, the measured pressure drops and friction factor of single phase oil and single phase water were compared with existing empirical relations and good agreement was noticed.

For a given flow rate (for all angles) the frictional pressure drop (FPD) has been found to increase from WC = 0 - 60 %. Further increase in WC causes FPD to decrease. For a given flow rate of 6000 BPD (θ = 15°), with increase in WC from 0 to 20, percentage increase in FPD is about 36%. Further increase from 15° to 30°, percentage increase in FPD has been found to be about 77%.

6. Nomenclature

A	Cross-sectional area of pipe $[m^2]$	ρ_w	Density of water [kg/m^3]
BPD	Barrel per day	ρ_o	Density of oil [kg/m^3]
Dh	Diameter of the pipe [m]	μ_o	Viscosity of oil [$Pa.s$]
f	Friction factor	μ_w	Viscosity of water [$Pa.s$]
ID	Inner diameter [m]	ΔP	Pressure drop [$inch\ H_2O, pa$]
L	Length of the pipe [m]	$\frac{\Delta P}{\Delta L}$	Pressure gradient $[\frac{Pa}{m}, \frac{inch\ H_2O}{m}]$
Re	Reynolds's number		

Acknowledgments

The authors acknowledge the support of Center for Engineering Research, Research Institute, King Fahd University of Petroleum and Minerals for this research work.

References

[1] Jing-yu Xu, Dong-hui Li, Jun Guo, Ying-xiang Wu (2010). Investigations of phase inversion and frictional pressure gradients in upward and downward oil–water flow in vertical pipes. *International Journal of Multiphase Flow*, v. 36, pp. 930–939.

[2] Pedram H., AlirezaH.i, Amir K (2015). Experimental investigation of oil–water two phase flow regime in an inclined pipe. *Journal of Petroleum Science and Engineering,* 136, pp. 12–22.

[3] Descamps M. N., Oliemans R. V. a., Ooms G., and Mudde R. F. (2007), Experimental Investigation Of Three-Phase Flow In A Vertical Pipe: Local Characteristics Of The Gas Phase For Gas-Lift Conditions, *Int. J. Multiph. Flow*, 33(11), pp. 1205–1221.

[4] Grassi B., Strazza D., and Poesio P. (2008), Experimental Validation Of Theoretical Models In Two-Phase High-Viscosity Ratio Liquid–Liquid Flows In Horizontal And Slightly Inclined Pipes, *Int. J. Multiph. Flow*, 34(10), pp. 950–965.

[5] Du M., Jin N.-D., Gao Z.-K., Wang Z.-Y., and Zhai L.-S. (2012), Flow Pattern And Water Holdup Measurements Of Vertical Upward Oil–Water Two-Phase Flow In Small Diameter Pipes, *Int. J. Multiph. Flow*, 41, pp. 91–105.

[6] Domenico S., B.Grassi., M. Demori, V. Ferrari, and P. Poesio (2011). Core-annular flow in horizontal and slightly inclined pipes: Existence, pressure drops, and hold-up. *Chemical Engineering Science*, v. 66, pp. 2853–2863.

[7] Karolina I., O.J. Nydal, and P. Angeli (2005). Phase inversion in dispersed liquid–liquid flows. *Experimental Thermal and Fluid Science,* v. 29, pp. 331–339.

[8] Dongjian Z., L. Guo, X. Hu, X. Zhang, and X.Wang (2006). Experimental study on local characteristics of oil–water dispersed flow in a vertical pipe. *International Journal of Multiphase Flow*, v. 32, pp. 1254–1268.

[9] Lum J.Y. L., T. Al-Wahaibi, and P. Angel (2006). Upward and downward inclination oil–water flows. *International Journal of Multiphase Flow*, v. 32, pp. 413–435.

IOP Conf. Series: Materials Science and Engineering **326** (2018) 012014 doi:10.1088/1757-899X/326/1/012014

A Review of Research on Bird Impacting on Jet Engines

Yuecheng Jin[1,*]

[1]The Sino-British College, University of Shanghai for Science and Technology, Shanghai, 200093

*SBC-15-1058@sbc-usst.edu.cn

Abstract. Bird strikes can lead to permanent deformations, sudden decrease of thrust, even engine failure during the flight. Bird strikes on rotating blades can also cause slices of birds hitting other parts which may lead to greater damages. Bird strikes cannot be completely avoided. However, reduction of bird impacting on jet engines can be achieved by suitable design and manufacturing, through the mathematical modelling, simulation analysis and practical experiment of jet engines.

1. Introduction

With Wright brothers operating the first fixed-wing aircraft off the ground in 1903, humans can fly like birds in the sky. According to Boeing's reports, three-quarters of bird strikes were related to aero-engines. Civil Aviation Authority (CAA) and International Civil Aviation Organization (ICAO) have similar conclusions. Bird strikes can lead to permanent deformations, sudden decrease of thrust, even engine failure during flight. Bird strikes on rotating blades can also cause slices of birds hitting other parts which may lead to greater damage. Bird strike cannot be completely avoided. However, reduction of bird impacting on jet engines can be achieved by suitable design and manufacturing, through the mathematical modelling, simulation analysis and practical experiment of jet engines.

The rest of the paper is organized as follows: numerical simulation analysis is illustrated in Section2, contents related to bird strike tests is carried out in Section3, anti-bird strike design of jet engines is given in Section 4. Conclusions are given in Section 5.

2. Numerical simulation analysis

Numerical simulation analysis relies on mathematics or computer-based methods, greatly reducing the cost and destruction of the experiments.

2. 1 Method of numerical simulation analysis

According to physical properties of the blades and bird during the impact, several transient states may exist, thus several possible methods as follows [1].

(1). Structure-structure coupling computation

The bird is considered as rigid body and the blade is considered as deformable target.

(2). Fluid-structure decoupled algorithm

The bird is considered as fluid during the impact and the blade is considered as deformed target. The bird and the blade response is modelled separately by building the bird impact model first then calculate the blade response.

(3). Fluid-structure coupling computation

The bird is considered as fluid during the impact and the blade is considered as deformed target. The bird and the engine is considered as the whole system during the impact.

(4). Contact-impact finite element method

The border of the bird and the blades is determined and the force during the impact is calculated by changing of momentum. This method can reduce the errors caused by the coupling computation and has higher precision.

2.2 Probability of bird impacting on different area of the engine blade

2.2.1 Probabilities of impacting regions on the blades. Ma Li et al. [2] defined the engine into three parts: blade apex, blade body and blade root. The probability of impact region that the bird will hit is dependent on the area of the three parts and it is defined as:

$$P_{1a} = \frac{S_a}{S_T}, P_{1b} = \frac{S_b}{S_T}, P_{1r} = \frac{S_r}{S_T} \tag{1}$$

Where Pa is the probability of the bird impacting on the blade apex, P_{1b} is the probability of the bird impacting on the blade body; P_r is the bird impacting on the blade root. S_a, S_b, S_r and S_T are the area of the blade apex, body, root and the total area, respectively [2].

2.2.2 Probabilities of bird passing the air inlet. According to Ref.[2], the passing probabilities is the ratio of the diameter of the bird model D and the width of air inlet interval L.

$$P_2 = \frac{L-D}{L} \times 100\% \tag{2}$$

The relationship between the width of the inlet and the radius of the bird model R is:

$$L = \alpha R - \delta \tag{3}$$

Where α is the angle between the two holder support, δ is the thickness of the holder support, R is the radius of the impacting point. The analysis only consider that the whole bird passing through the air inlet, in real situations that part of the birds tissues may pass through and increase the possibilities.

2.2.3 Probabilities of impacting regions on the blades. Considering all the probabilities, the equations can be get as $P=P_1 \times P_2$, and the probability that the bird will most likely to be hit is the apex blade part and the root is less likely to be hit. This conclusion is applicable to almost all jet engines.

2.3 Bird modeling

2.3.1 Smoothed Particle Hydrodynamics method (SPH method). Barber et al.[3] firstly demonstrated that the material constitution birds can be modelled as fluid during high velocity impact with rigid body because the birds' tissue strength is significantly lower than the impacting stress. The SPH method[4] is used to simulate bird model is based on normalized corrected Kernel interpolation which is first order consistent and conservative.

2.3.2 Bird geometry. Budgey [5] proposed the relationship between bird mass and density, bird diameter and mass. Budgey also proposed relationship between the mass of the bird model and slicing effect in his report.

$$\rho = -0.063 \times \log_{10} m + 1.148, \log_{10} D = 0.335 \times \log_{10} m + 0.900 \tag{4}$$

Where ρ denotes density, m is the mass, D represents the bird diameter.

Currently, three main shapes of the bird model are used in the strike tests and simulations which are the hemispherical-ended cylinder, ellipsoidal and flat-ended cylinder. The hemispherical-ended cylinder is the most commonly used shape in the test and stimulation which is adapted by McCarthy et al.[6], Langrand et al.[7] and Airoldi and Cacchione[8][9]. The ellipsoidal is important because it is recommended by IBRG.

2.3.3 Bird material. As mentioned above, the strength of birds is significantly lower than the strength of the blades, birds are considered as continuous fluid. The most commonly used bird material model is the Isotropic-Elastic-Plastic-Hydrodynamic(IEPH) material proposed by Anghileri et al.[10], which takes the slicing effect of the artificial bird into consideration. The forces acting on the blade can be calculated using the impulse which is the change of momentum. The equation of state (EOS) with the IEPH material model is applied to accurately model the hydrodynamic response [6].

$$P = P_0 + B\left[\left(\frac{\rho}{\rho_0}\right)^{\gamma} - 1\right] \tag{5}$$

Where P_0 is the reference pressure, B and γ are experimental parameters which are dependent on the bird model.

2.4 Fan blade modeling

Since each engine has different geometry and material model. Three different engines are taken as the examples in this section.

2.4.1 Engine 1[11]. As shown in Figure 1, the engine consists of 22 equally distributed engine blades attached to the disks with 16.36 degrees between the two following blades. This disk part is considered as rigid body, neglecting the impacting effect. Corresponding radiuses are measured from the axis of rotation of the rotor. The external radius is 570mm measured from the center of rotation to the tip of the blade. The impacting radius is the radius of the impacting point which is 514mm. The internal radius is the radius from the axis to the root of the blades which is 175mm.

The blade and its support were meshed using TrueGrid software. The blade was meshed with 21,400 elements: 107 elements along lengthwise, 50 elements width wise and 4 elements through the thickness.

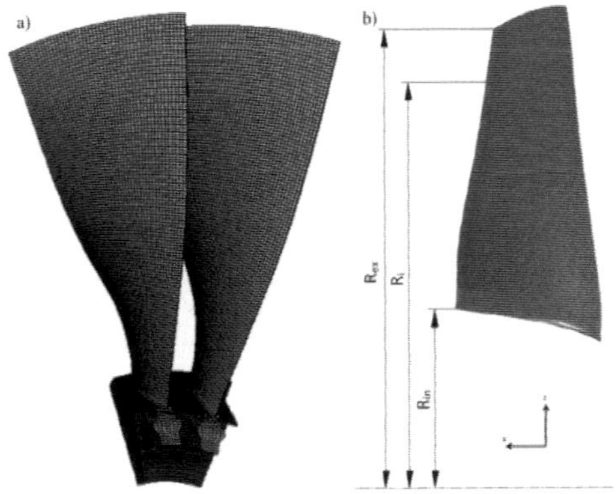

Figure 1. (a).Finite element part of the model,(b)Blade dimension measured form the leading edge.[11]

2.4.2 Engine 2[12]. As shown in Figure 2, a typical modern jet engine fan blade was modeled. The fan blade assemble consists of 18 equally distributed blade sector and a hub section. Only two blades is collected in the simulation in order to increase the efficiency of calculation. The mesh has 69,470 elements and 127,972 nodes. The engine blades are typically made of titanium alloy Ti-6Al-4V and the material model used in the simulation is Johnson-Cook viscoelasticity material model.

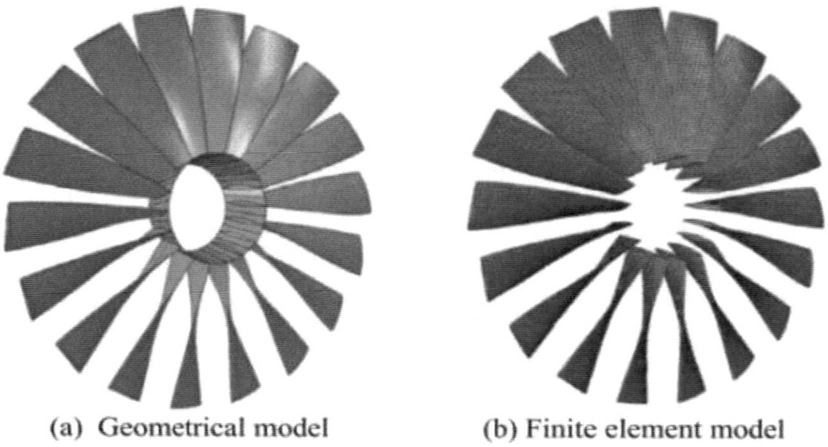

Figure 2. Geometrical model and finite element model [2]

2.4.3 Engine 3[13]. As shown in Figure 3, the hub is assumed be rigid body compared with the blade. The inner and outer radius of the blades is 0.51m and 1.47m, respectively. Shell-163 elements in LS-DYNA were used to simulate the blade with 31 nodes in the axial direction and 61 nodes in the radial direction. The thickness of the blades can be presented as:

$$\Delta = 0.026 - 0.00463y^2 - (z + 0.1)^2 \tag{6}$$

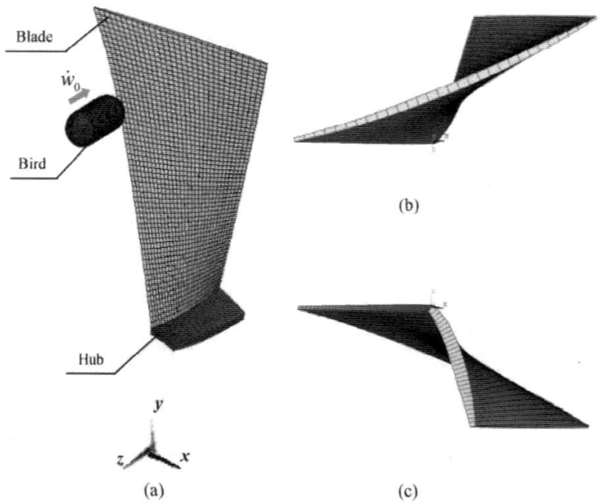

Figure 3. Distributed 3D fan blade finite-element model[13]

2.5 Simulation process and results

2.5.1 Simulation process. According to the simulations, there are also common points that the bird impacts the engine blades in the analysis. The bird impacting the engine blade can be divided into three stages.

In the first stage, the bird model strikes the front edge of the bird with large relative velocity. In this stage, the blade cutting the bird model and under the great impacting force, the blade bend to the other side and the front edges shows evident plastic deformation. There is concave edges occurs at the impacting point and cracks and off-falling emerges. This can cause further damage when the engine continues working with cracks propagation and metallic failure which is called the secondary damage.

In the second stage, the bird slices cutting by the blade hitting the blade causing the whole blade to bend over. Plastic deformations can occur at the shrouds and the roots and the damage is smaller compared to the first stage. In the third stage, the blade oscillate and recover because its own elasticity. This can cause further crack propagation, off-fall and fatigue.

2.5.2 Bird shape influence. The impact location was set at the radius of 514mm (86% of the blade span) from the rotation axis. Two different bird model, hemispherical bird model and ellipsoidal bird were selected in the simulation to show the difference of the impact. All the other parameter was controlled to show the direct comparison [11].

(1) Hemispherical bird model

For the hemispherical bird model, the stress generated by the impact distributed over a larger area than the ellipsoidal bird model. For the hemispherical bird, deformation of the blade was more widely spread, and the plastic deformations were observed at the trailing edge of the impacted blade and the root of the leading edge of the leading blade.

(2) Ellipsoidal bird model

The ellipsoidal bird model has the highest plastic strain observed closing to the impacting point. The ellipsoidal bird impact the leading blade suffered only local deformation (deformation around the impact point). The ellipsoidal bird model has the highest plastic strain observed closing to the impacting point.

2.5.3 Bird incidence angle. According to the simulation, the normal incidence can lead to the maximum impacting force and plastic strains causing great damage when the other parameters are controlled. When the incidence angle is greater or equal to 60 degrees, the impacting force is distinctly reduced and the blade will only suffer from elastic strain [13].

2.5.4 Bird impact location. The response of the blades was strongly dependent on the bird impact locations. The impact location moves towards the rotational axis the contact forces were reduced. The contact force magnitude was related to the blade pitch angle which increases with the distance from the rotation axis. Consequently, this influenced the bird slice size and its change of momentum during the impact [2], as shown in Table 1.

Table 1. Damage caused by bird impact on different location

Bird strike location	Blade damage degree
Fan blade root	Plastic deformation in the edges of the root, little deformations in the fan blade body and apex, no cracks and off-falling.
Fan blade body	Big plastic deformation at the front edge of the blade body, deformations found at the root and apex of the fan blade with off-falling and relatively big deformation at the fan blade body.
Fan blade apex	Great deformations at the front edges of the blade apex with off-falling, plastic deformation at the body and root of the fan blade with cracks. Big plastic deformation at the fan blade body.

3. Bird impacting experiment

3.1 Experiment process

The bird model is prepared according to simulation and the velocity of the bird and rotational speed is according to the international military standard. The impact location is chosen at the apex of the blade since it can cause the greatest damage in the simulation. The experiment is mainly to find real impacting stress and force and the whole experiment is recorded by high-speed camera.

In all the experiments, the bird model hit disintegrates into slices and there is distinct bending to the other side of the blade at the striking point. The bird model continues hitting the following blade with the similar phenomenon but smaller concave edges at the impacting point until the bird model is consumed totally [14].

3.2 Damage classification and standard

According to the American aviation turbine bird injecting statistical report [15], the damage is classified into 6 sorts. According to the Rolls-Royce bird injecting standard [16], the damage is classified into 5 sorts, as shown in Table 2.

Table 2. Damage Classification of Fan Blade Bird Strike by R.R.

Category	**Description**
Tip bending	Caused by the strike of the blade tip, resulting in loss of thrust
Front chord dent	The degree in loss of thrust depends on the dent depth
Axial tearing	Crack propagation might result in chipping
Radial tear at the boss root	Chipping due to crack propagation at the boss root of the blade with shrouds
Breakage at the blade root	The whole blade flies out

The Rolls-Royce also divided the damage into 6 levels[16], as shown in Table 3.

Table 3. Blade Damage Grade of Bird Ingestion Test by R.R.

Damage Grade	Damage condition
2	Front, rear chord bending or warping
2	Boss bumping and chipping
4	Slight blade blending
4	Several local warping
6	Bending of severe plastic deformation on the impact
6	radius
8	Deformed boss, but still abutted
8	Bending of severe plastic deformation at the blade root
8	The boss is not abutted after the test
10	Scraping with other blades, or being cut off
15	Tip fracture
	Blade fracture from the blade root

3.3. The definition of damage parameter and evaluation method

3.3.1 The equivalent plastic strain damage parameter. The blade failure modes such as the overall deformation, curling, dent, tear and breakage are caused by the plastic strain of the local or root material of the blade. According to the uniaxial tension curve in the blade material property, the plastic strain of the blade fracture and failure can be determined. The maximum equivalent plastic strain when the blade is struck by the bird can be obtained through the numerical calculation or test of the impact response. ή, the plastic strain damage parameter can be defined as[17][18]:

$$\eta = \frac{\varepsilon_{p,\max}}{\varepsilon_p^*} \tag{7}$$

3.3.2 Blade deformation parameter. Due to the lateral vibration and torsional movement after the blade is struck by the bird, the tip may be exposed to the adjacent blades, or collide with the casing, which will lead to mechanical failures of the engine. According to the actual blade design, the blade tip displacement can be determined when the blade fails, and the actual maximum displacement of the blade tip can be obtained through the numerical calculation or test of the impact response. λ, the blade deformation damage parameter can be defined as[18]:

$$\lambda = \frac{U_{Tip,\max}}{U^{*}_{Tip}} \qquad \text{`(8)}$$

After calculation, if $\eta \geq 1 \text{ or } \lambda \geq 1$, then the blade is considered to be metallic failure. If the damage parameter is $0 \leq \eta < 1 \ or \ 0 \leq \lambda < 1$, then the level of destruction and the corresponding risks can be evaluated.

3.4 Experimental results and comparison

The experiment shows the similar phenomenon to the simulation results. The bird impacting can be divided into three stages. The bird impacts the front edges of the blades in the first stages with the biggest impacting force causing plastic deformation at the front edges and shrouds. This is the most dangerous periods. The second part is the slices of birds hitting the blades with relatively small impacting force and causing little damage. The third period is the blades recovering under its own elasticity and vibrating which does not cause new damage but has the possibility to aggravate the formal damage.

The experimental result and the simulation show the similar trend of bird impacting on the blades and the same damage sort at each stage. The difference is the specific value calculated at the instant time which may cause by systemic and random errors.

4. Designs for reducing damage of bird impacting

To reduce the damage of the bird strike to the engine blades, several designs are proposed. After searching, the designs can be divided into two sorts: one is using technical method to lower the possibilities of bird from attacking the engine blades as results reducing the damage indirectly; the other method is applying new designs to reduce the impact damage directly.

4.1 Indirect design

4.1.1 Changing inlet area. According to some research, several Russia jet fighters will close the major inlet and use vice inlet during the landing process. This method is to reduce windward area of the engine blade thus reducing the possibilities of the bird impact.

4.1.2White flash lines on engine cowling. Researchers have also presented other methods. For example, people now paint a white spiral of a gradually varied width on the conic engine cowling which rotates at a high speed with the rotor of the fan. When the engine is running, these spirals appear as a white flash circle which discourages birds from flying into the engine. Although these measures work to some extent, they are far from perfect.

4.2 Direct design

4.2.1. Wide-chord hollow fan blades. Wide-chord hollow fan blades first launched by Rolls-Royce features strong intensity and their weight is much less than that of narrow-chord solid fan blades which not only increases efficiency of engines but also reduces vibratory fatigue of engines and improves anti-bird-strike performance. A number of achievements with regard to bird-resistant fan blades have been made based on many of the above theories. Pratt & Whitney Group has developed a kind of milling groove wide-chord hollow fan blades, improving its anti-strike performance. Fan blades of such structure are not only applied to PW4084 engines of Boeing777 but also to F119 of F22, fourth-generation fighters [19].

4.2.2 Prevent bird strike with blast wave. The idea of preventing bird strike with blast waves presented is based on military helicopter using reactive oriented explosive attaching on the propeller to destroy the inevitable barrier and protect the blades. The specific design involves attaching a certain amount of linear oriented explosive to inlet fan blades of jet engines. When a bird hits the blade, the explosive explodes, shattering bird carcass into pieces. As only a small amount of oriented explosive is applied to shatter bird carcass, it will not cause damage to fan blades and can ensure that blast waves are controlled within the designing limits of jet engines. The article herein is going to calculate the impact of such blast waves on the stability margin of jet engines and make sure the blast waves are within the designing range of the engine while managing to smash bird carcass [19].

4.2.3 Protection cables. The specific designing idea of a bird strike protection device is to attach a certain number of protection cables to the front inlet fan hub of jet engines. When the inlet fans are working, protection cables rotate at an angular velocity, creating a protective shield in front of the engine which is able to smash bird carcass before it hits the engine so as to reduce damage to engines by bird strikes. It is primarily decided to make protection cables from composite materials which feature robust impact strength to ensure the cables will not break when smashing bird carcass [20].

5. Conclusions

In this article, the research status of the bird impacting on the jet engine blades was stated. Firstly, the bird impacting research should include the establish of the standardized bird models which can give comparable experiment results and numerical simulations. Secondly, the development numerical simulation method and model like the SPH method and using finite element method is the top priority in the bird impacting research to visualize the complex process and reduce the experimental expense. Then the damage level is determined according to the experiment. Finally, several possible anti-bird strike devices is introduced.

The future anti-bird strike method can be concentrated on the bird incidence angle since the change of the impact can greatly reduce the impact damage. Considering the backswept wings can dramatically reduce the air friction at high speed, a conjecture of backswept wide-chord hollow fan blades jet engine blade may be designed to promote the air inflow efficiency and reduce frictional energy loss as well as changing the bird impact angle thus greatly promote the anti-bird strike ability.

References

[1] Chen Wei, Guan Yupu, Gao Deping. Numerical simulation of the transient response of blade due to bird impace [J]. Chinese Journal of Aeronautics, 2003, 24(6):531-533.
[2] Ma Li, Jiang Jiayu, Xue Qingzeng. Research on bird impact of aeroengine first stage fan[J]. Aeroengine, 2014, 40(2):65-69.
[3] Barber J P, Fry P F, Klyce J M, et al. Impact of Soft Bodies on Jet Engine Fan Blades[J]. Impact of Soft Bodies on Jet Engine Fan Blades, 1977.
[4] Chevrolet D, Audic S, Bonini J. Bird Impact Analysis on a Bladed Disk[J]. Bird Impact Analysis on A Bladed Disk, 2003.
[5] Budgey R. The development of a substitute artificial bird by the international Bird strike Research Group for use in aircraft component testing. International Bird Strike Committee ISBC25/WP-IE3, Amsterdam[C]// International Bird Strike Committee. 2000.
[6] M A McCarthy, J R Xiao, C T McCarthy, et al. Modelling bird impacts on an aircraft wing – Part 2: Modelling the impact with an SPH bird model[J]. International Journal of Crashworthiness, 2005, 10(1):51-59.
[7] B Langrand, AS Bayart, Y Chauveau, et al. Assessment of multi-physics FE methods for bird strike modelling-Application to a metallic riveted airframe[J]. International Journal of Crashworthiness, 2002, 7(4):415-428.
[8] Airoldi A, Cacchione B. Numerical analysis of bird impact on aircraft structures undergoing large deformations and localised failure. Impact Loading of Lightweight Structures 2005;49.
[9] D.H. Zhang, Q.G. Fei, H. Liu, Effects of bird's striking orientation on bird impact analysis

based on a realistic bird model, J. Vib. Shock 34(22) (2015) 103–108.
[10] Anghileri M, Sala G. Theoretical assessment, numerical simulation and comparison with tests of birdstrike on deformable structures. Proceedings of the 20th ICAS congress, Sorrento, Italy; September 8e13, 1996. p. 665e74.
[11] Vignjevic R, Orłowski M, Vuyst T D, et al. A parametric study of bird strike on engine blades[J]. International Journal of Impact Engineering, 2013, 60(60):44-57.
[12] Zhang D, Fei Q. Effect of bird geometry and impact orientation in bird striking on a rotary jet-engine fan analysis using SPH method[J]. Aerospace Science & Technology, 2016, 54:320-329.
[13] R. H. Mao, S. A. Meguid, T. Y. Ng. Effects of incidence angle in bird strike on integrity of aero-engine fan blade [J]. International Journal of Crashworthiness, 2009, 14(4):295-308.
[14] Zhang Haiyang, Yu Duokui, Wang Xiangping. Numerical and experimental investigation of damage of bird impact on fan blades [J]. Journal of propulsion technology, 2015, 36(9):1382-1388.
[15] Engine, aircraft, turbojet and turbofan, general specification for, MIL-E5007D.
[16] Horsley. J. The rolls-royce w ay of validating fanintegrity. AIAA 93-2602.
[17] Huang Zhiyong, Chen Wei, Zhao Haiou. Parameters and methods for evaluating bird impact damage of fan/compressor blade[J]. Aeroengine, 2005, 31(1):28-30.
[18] Zhiyong Huang, Wei Chen, Haiou Zhao,et al. Evaluation of bird impact parameter and method [J]. Aviation engine, 2005, 31(1):28-30.
[19] Peifeng Wen, Shuanqin Cao. Problem of aviation engine and its solution[J]. International Aviation, 1996(10):50-51.
[20] Hou Bingduo. Turbofan engine bird collision protection device design and simulation research[D]. Nanchang Hangkong University, 2012.

IOP Conf. Series: Materials Science and Engineering **326** (2018) 012015 doi:10.1088/1757-899X/326/1/012015

Numerical Analysis of the Cavity Flow subjected to Passive Controls Techniques

Kursad Melih Guleren[1, a], Seyfettin Turk[1, b], Osman Mirza Demircan[2, c] and Oguzhan Demir[2, d],

[1]Anadolu University, Faculty of Aeronautics and Astronautics, 26470, Eskisehir, TURKEY
[2]Istanbul Technical University, Faculty of Aeronautics and Astronautics, 34469, Istanbul, TURKEY

E-mail: [a]kmguleren@anadolu.edu.tr; [b]seyfettinturk@anadolu.edu.tr; [c]demircano16@itu.edu.tr; [d]demirog16@itu.edu.tr

Abstract. Open-source flow solvers are getting more and more popular for the analysis of challenging flow problems in aeronautical and mechanical engineering applications. They are offered under the GNU General Public License and can be run, examined, shared and modified according to user's requirements. SU^2 and OpenFOAM are the two most popular open-source solvers in Computational Fluid Dynamics (CFD) community. In the present study, some passive control methods on the high-speed cavity flows are numerically simulated using these open-source flow solvers along with one commercial flow solver called ANSYS/Fluent. The results are compared with the available experimental data. The solver SU^2 are seen to predict satisfactory the mean streamline velocity but not turbulent kinetic energy and overall averaged sound pressure level (OASPL). Whereas OpenFOAM predicts all these parameters nearly as the same levels of ANSYS/Fluent.

1. Introduction

Cavity flows are encountered frequently in aeronautical and mechanical engineering applications. Studies related to high-speed cavity flows date back to 1960s with the pioneering experimental study [1] by which the pressure oscillation modes were analysed. Later on, at subsonic and transonic regimes the cavity flows were classified as open, transitional and close in [2]. The flow is immediately separated from the leading edge of the cavity similar to the flow behaviour of backward facing step. Open cavity flow is maintained when the high shear emerged after the separation meets the trailing edge following a hypothetical straight line. Transitional flow occurs when the flow penetrates the cavity but it is not attached on the cavity floor. As for the closed cavity flow, the flow attaches the cavity floor but separates again to reach the leading edge of the cavity. The control of the cavity flow is utmost important to decrease the cavity noise. One of the most cited study on the passive control of the cavity flow is known by the M219 case [3]. Here the Mach number is kept constant at 0.85 for a cavity of L/D ratio of 5 and W/D ratio of 1. Five passive control cases are available: M219 cavity slanted wall (M219-SW), M219 transverse rod (M219-TR), M219 1δ flat-top spoiler (M219-FTS) where δ is the boundary layer height, M219 2δ saw-tooth spoiler Type 1(M219-STS1) and Type 2 (M219-STS2). A schematic illustration of the clean cavity and the adopted passive control models are given in Figure 3 and 4, respectively.

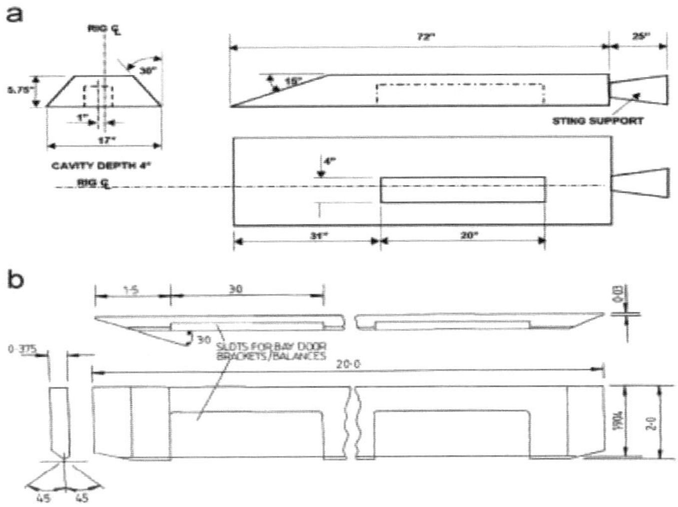

Figure 1. M219 experimental cavity geometry [3]

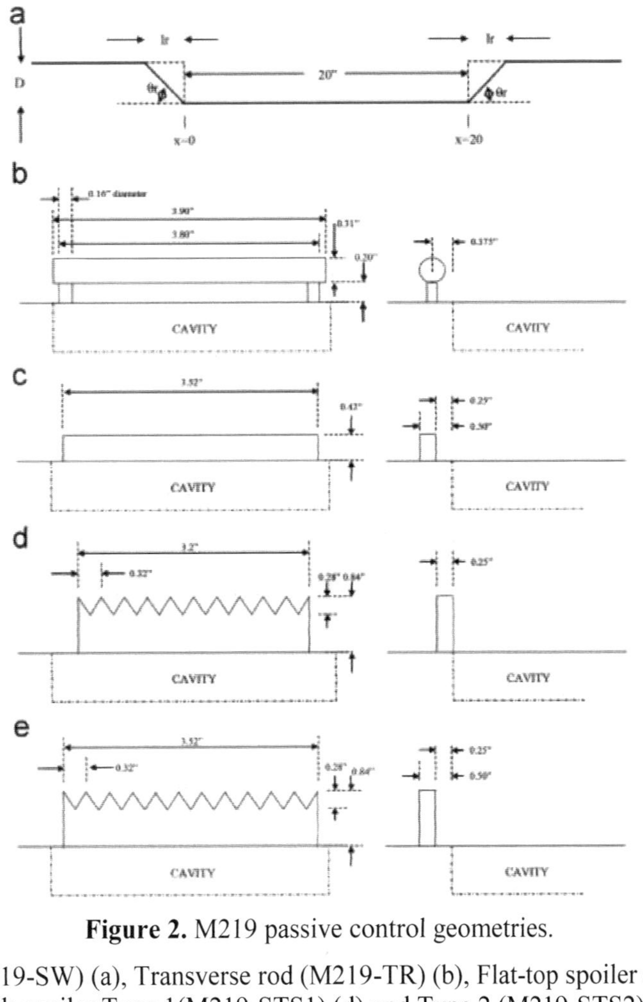

Figure 2. M219 passive control geometries.

Slanted wall (M219-SW) (a), Transverse rod (M219-TR) (b), Flat-top spoiler (M219-FTS) (c), Saw-tooth spoiler Type 1(M219-STS1) (d) and Type 2 (M219-STS2) (e) [3].

The aim in this study to conduct numerical analysis of the turbulent flow subjected to the M219 geometries seen in Figures 1 and 2 with the available commercial and open-source flow solvers and compare the results with the available experimental [3] and LES data [4]. Firstly, mean streamwise velocity and turbulent kinetic energy are compared for the clean cavity (M219-CC). Then, OASPL data are compared for M219-CC, M219-SW, M219-TR and M219-STS2.

2. Methodology

The fluid domain for each analysis is constructed using the sketches of the related experimental cavity models. Computational domain and the boundary conditions adopted for this domain is shown in Figure 3 and Table 1, respectively. Free stream Mach number and temperature is kept constant at 0.85 and 288 K, respectively. The same number of mesh (around 600 000) is used for RANS and URANS cases. For LES, around 6 million mesh number is employed. Pressure based solution is preferred for ANSYS/Fluent software for LES and RANS (k-w SST) cases. For OpenFOAM URANS solutions, the sonicFoam module is used. As for the SU2 RANS cases, the methodology explained in [5] is preferred.

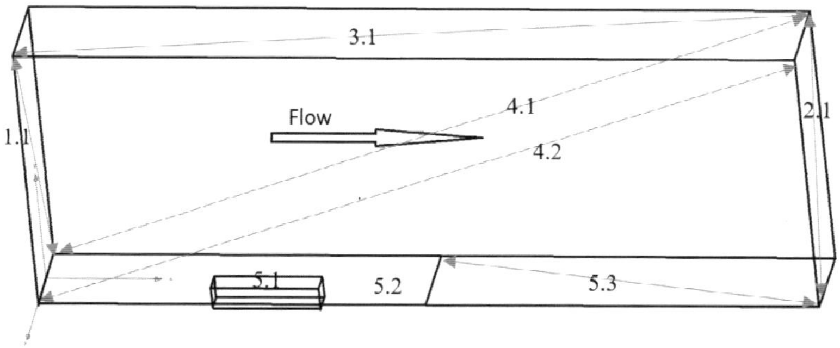

Figure 3. Computational domain for M219 flow analysis

Table 1. Formatting sections, subsections and subsubsections

#	Boundary	Dimension	Applied Boundary Condition		
			ANSYS Fluent	SU^2	OpenFOAM
1.1	Inlet	0.43×1.27	Farfield	Farfield	Wave transmissive
2.1	Outlet	0.43×1.27	Outlet	Outlet	Wave transmissive
3.1	Top	$\mathbf{0.43 \times 3.66}$			
4.1	Side left	$\mathbf{1.27 \times 3.66}$	Farfield	Farfield	Wave transmissive
4.2	Side right	$\mathbf{1.27 \times 3.66}$			
5.1	Cavity	$\mathbf{0.1 \times 0.1 \times 0.5}$			
5.2	Cavity near	$\mathbf{0.43 \times 1.83}$	Adiabatic, no-slip	Adiabatic, no-slip	Adiabatic, no-slip
5.2	Cavity downstream	$\mathbf{0.43 \times 1.83}$			

3. Results

Firstly, mean streamwise velocity and turbulent kinetic energy predictions are shown in cavity regions as reported in Figures 4 and 5. For the comparisons with the current predictions, LES data of [3] is used. To validate these LES data, LES of ANSYS/Fluent is performed for M219-CC case. As seen our LES predictions are in good agreement with LES predictions of [3]. OpenFOAM streamwise velocity results are very promising except the profile at x/L=4 where the streamwise velocity is underpredicted at -0.5 < y/D < 0.5. There is also a small zigzag region which is suspected to occur due to some numerical deficit which could be related to the data sampling of streamwise velocity. SU^2 predictions agree well with LES of [3] apart from some small underpredictions at x/L=3 and 4. Considering the predictions of OpenFOAM URANS and ANSYS/Fluent RANS, one can say that SU^2 is superior to both flow solvers in calculation of the streamwise velocity. However, this situation is not seen in Figure 4. SU^2 underpredicts the turbulent kinetic energy significantly at x/L=3 and 4 where ANSYS/Fluent RANS performs better than those two open-source flow solvers. However, noticeable underpredictions and over predictions exist at x/L=2. OpenFOAM URANS predictions agree very well with LES predictions for the first half of the cavity, but later they are overpredicted close to the trailing edge of the cavity. Considering both the trend and accuracy of the turbulent kinetic energy results, once can conclude the OpenFOAM is superior over ANSYS/Fluent and SU^2.

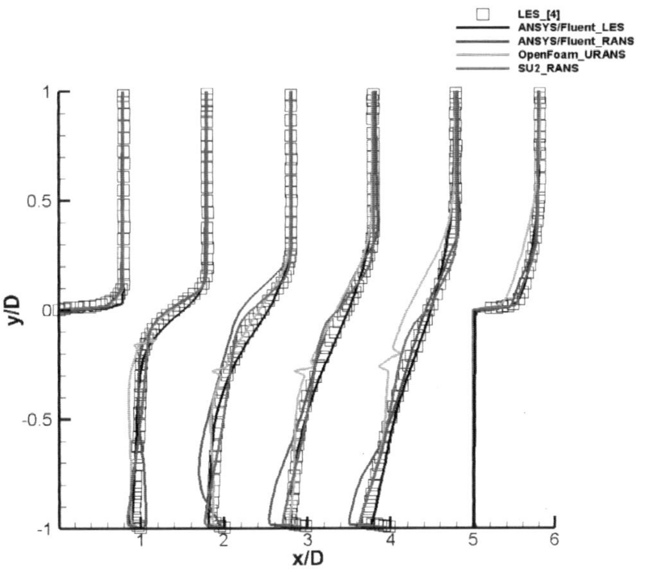

Figure 4. Streamwise mean velocity predictions (U_x/U_∞)

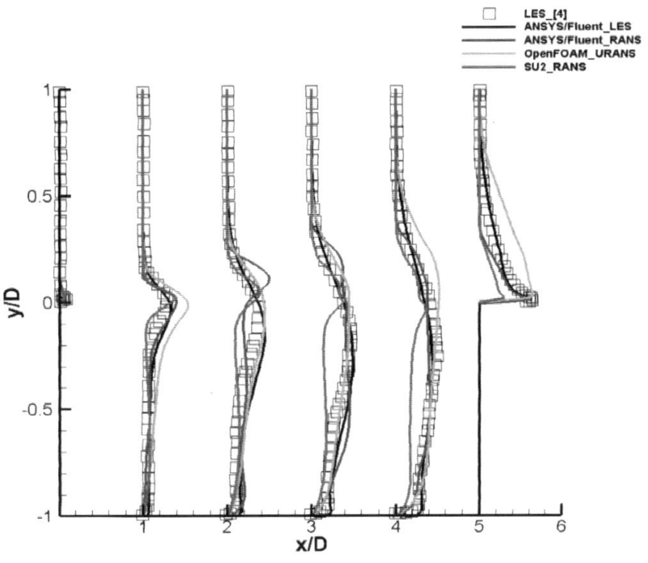

Figure 5. Turbulent kinetic energy predictions (k/ U_∞^2)

The success of a passive control method lies on the fact that the sound pressure level associated with the cavity surface decreases in comparison to that of the clean cavity where no passive control exists. Figure 6 shows the overall-average sound pressure level (OASPL) on the mid-line of the cavity floor for the clean cavity (Figure 6a) and for the passive control methods (Figure 6b,6c and 6d). OpenFOAM and SU^2 URANS as well as ANSYS/Fluent LES OASPL data are compared with previous experimental OASPL data [3]. Current LES overpredicts OASPL data for all cavity simulations. Nevertheless, the monotonic increase from the cavity leading edge to the cavity trailing edge reported by the experiment is satisfactory captured by our LES. OpenFOAM predictions show a very close trend with LES prediction for clean cavity case but indicate underpredictions and overpredictions for the passive control cases. Therefore, one can say that OpenFOAM predicts accurately but not precisely. It is believed high precision can be obtained by increasing the number of mesh and it will be in the plans of future work. Unlike the good performance of SU^2 for velocity and turbulence kinetic energy, its performance in OASPL predictions is truly disappointing. In fact, SU^2 underpredictions in OASPL reach sometimes 30 dB. These underpredictions are attributed to the available wall-functions in the code. It is not shown here but it was seen that decreasing the y^+ around 10 improves OASPL results significantly.

4. Conclusion and Future Work

In the present work, the most common open-source flow solvers OpenFOAM ve SU^2 are tested for the flow predictions of high speed cavity flows by comparing the available experimental and LES data. A relatively coarse grid (around 600 000) with high y^+ values (~100-300) are deliberately chosen for both flow solvers. Considering the predictions of velocity, turbulent kinetic energy and sound pressure levels, OpenFOAM is seen to perform satisfactorily and can be used as a design tool of passive and active control methods to alleviate the sound caused by the high-speed cavity flows.

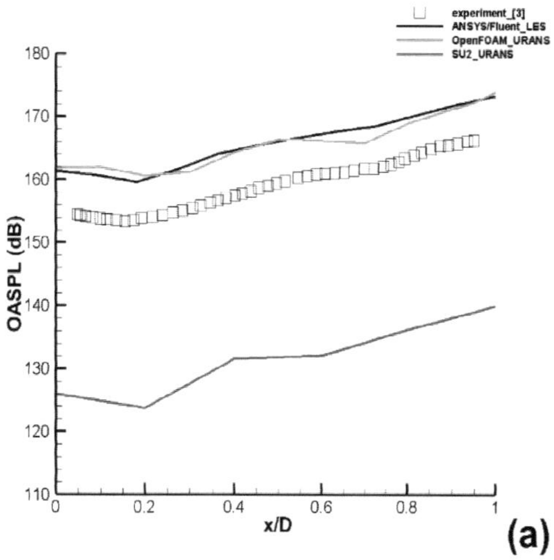

(a)

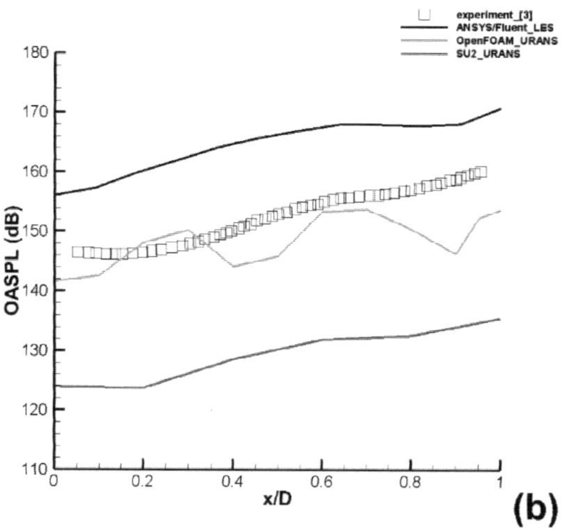

(b)

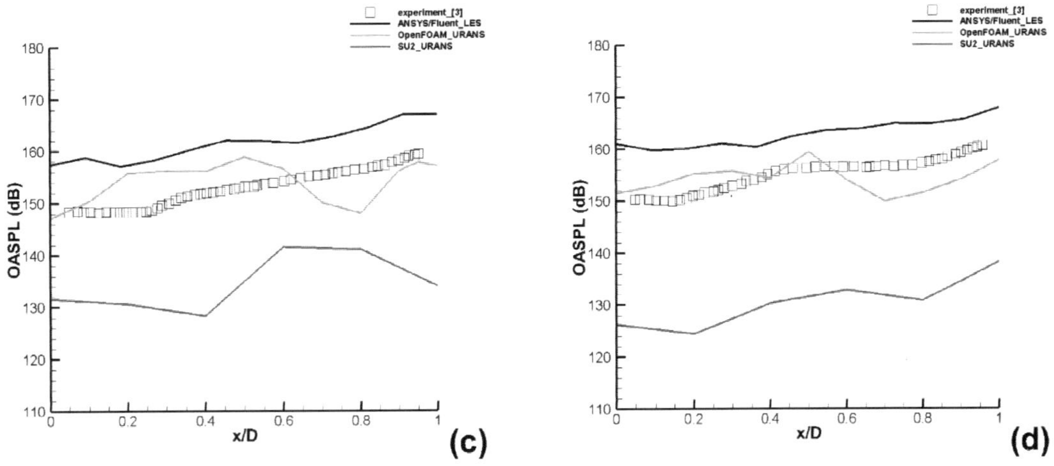

Figure 6. OASPL predictions for M219-CC (a), M219-SW (b), M219-TR (c), M219-STS (d)

Acknowledgement

Numerical analyses in this study were run on the National Center for High Speed Computing UHEM (**http://en.uhem.itu.edu.tr/**). This study is supported by the Turkish Aerospace Industries (**https://www.tai.com.tr/en**) under the grant of TM-2031. The authors are grateful to the authorized staff for their support and assistance with this study.

References

[1] Rossiter J E 1964 *Wind tunnel experiments on the flow over rectangular cavities at subsonic and transonic speeds*. Ministry of Aviation; Royal Aircraft Establishment; RAE Farnborough.

[2] Plentovich E B, Stallings Jr R L and Tracy M B 1993 *Experimental cavity pressure measurements at subsonic and transonic speeds* Technical Paper 3358 NASA

[3] Nightingale D, Ross J and Foster G 2005 *Cavity unsteady pressure measurements— examples from wind-tunnel tests* Technical Report Version 3 Aerodynamics & Aeromechanics Systems Group QinetiQ

[4] Larcheveque L, Sagaut P, Le Thien-Hiep and Comte P 2004 *Large eddy simulation of a compressible flow in a three-dimensional open cavity at high Reynolds number* Journal of Fluid Mechanics vol 516 p 265–301

[5] Economon T D, Palacios F, Copeland S R, Lukaczyk T W and Alonso J J 2016 *SU2: An Open-Source Suite for Multiphysics Simulation and Design* AIAA Journal Vol. 54 No. 3 p. 828-846

IOP Conf. Series: Materials Science and Engineering **326** (2018) 012016 doi:10.1088/1757-899X/326/1/012016

Linear Strength Vortex Panel Method for NACA 4412 Airfoil

Han Liu

8989 West Rd., Houston, TX 77064

hanliu@umich.edu

Abstract. The objective of this article is to formulate numerical models for two-dimensional potential flow over the NACA 4412 Airfoil using linear vortex panel methods. By satisfying the no penetration boundary condition and Kutta condition, the circulation density on each boundary points (end point of every panel) are obtained and according to which, surface pressure distribution and lift coefficients of the airfoil are predicted and validated by Xfoil, an interactive program for the design and analysis of airfoil. The sensitivity of results to the number of panels is also investigated in the end, which shows that the results are sensitive to the number of panels when panel number ranges from 10 to 160. With the increasing panel number (N>160), the results become relatively insensitive to it.

1. Introduction

The development of airfoil theory to predict lift and pressure estimates for a given airfoil has gone through several stages. The first successful airfoil theory, which based on conformal transformation, was developed by Joukowski [1]. He represented a potential flow by a complex potential and maps the complex potential flow around the circle in ζ plane to the corresponding flow around the airfoil in the z plane, which makes it possible to use the results for the cylinder with circulation to calculate the flow around an airfoil. However, it can only apply to a particular family of airfoil shape and all members of this family have a cusped trailing edge, which is inconsistent with the practical situation that has trailing edges with finite angles. Though Karman and Trefftz devised a more general conformal transformation giving the airfoils with trailing edges of finite angle, it still cannot be widely used in aerodynamic design since this technique cannot be extended to three-dimensional or high-speed flows. The second airfoil theory is the thin airfoil theory. In thin airfoil theory, the airfoil is replaced with its mean camber line. The flow pattern is built up by placing a bound vortex sheet on the camber line and adjusting its strength so that the camber line becomes a streamline of the flow. Within this framework, the theory adequately predicts lift and moment for thin airfoil. Nevertheless, its drawback is also obvious – it cannot be applied to arbitrarily thick airfoils because of the ignored thickness effects [2]. With the advent of digital computers offers the attractive alternative of a numerical rather than an analytical solution, a new method in aerodynamic design is widely used nowadays – the panel method. It relies on the distribution of singularities on discrete segments of the airfoil surface. By satisfying no penetration condition and Kutta condition, a system of linear algebraic equations to be solved for the unknown singularity-strength is created, with which, the lift coefficients and pressure distribution can be easily predicted. Panel method can be applied to airfoil section with any thickness and camber [3]. Computational Fluid Dynamics (CFD) is another option to provide accurate and reliable results but is more computationally expensive and complicated than panel methods. For preliminary stage, panel methods are accurate and less time consuming. Therefore, due to these advantages, it is necessary to investigate its principle for future use.

Panel methods have one key feature that distinguishes themselves from each other: the type of singularity element used to describe the flow field around the airfoil: a source, doublet or vortex element. As vortex element can model both pressure and lift, this article focuses on formulating numerical models for two-dimensional potential flow over the NACA 4412 Airfoil using linear vortex panel methods in the Matlab environment. By satisfying the no penetration boundary condition and Kutta condition, the circulation density on each boundary points (end point of every panel) are obtained and according to which, surface pressure distribution and lift coefficients of the airfoil are predicted and validated by Xfoil, an interactive program for the design and analysis of airfoil created by MIT. The sensitivity of results to the number of panels is also investigated in the end, which shows that the results are sensitive to the number of panels when panel number ranges from 10 to 160. With the increasing panel number (N>160), the results become relatively insensitive to it.

2. Methodology

For incompressible, inviscous and irrotational flow, the vector velocity can be represented as the gradient of a scalar velocity potential, $\vec{V} = \nabla\phi$, and the resulting flow is referred to as potential flow. According to continuity equation $\nabla \cdot \vec{V} = 0$, velocity potential satisfies the Laplace's equation $\nabla^2\phi = 0$. It is a liner partial differential equation and can be solved subject to no penetration boundary condition that no flow can cross the surface of the object [4].

2.1. To generate grid and establish NACA 4412 Airfoil geometry

a) Grid generation

Cosine Spacing can be used as the discretization schemes of airfoil. Its expression

$$\frac{x}{c} = \frac{1}{2}(\cos\zeta + 1) \ , \zeta = 0 - 2\pi \quad (1)$$

This scheme provides a fine discretization at the leading and trailing edge of the airfoil.

b) Establishment of airfoil geometry

The NACA 4-digit series is defined by four digits. First digit describing maximum camber as percentage of the chord. Second digit describing the distance of maximum camber from the airfoil leading edge in tens of percents of the chord. Last two digits describing maximum thickness of the airfoil as percent of the chord. E.g. NACA 4412, m=4%, p=40%, t=12%. The formula for the shape of a NACA 4-digit airfoil can be given by:

$$y_t = \frac{t}{0.2}(0.2969\sqrt{x} - 0.126x - 0.3516x^2 + 0.2843x^3 - 0.1015x^4) \quad (2)$$

$$y_c = \frac{m}{p^2}(2px - x^2) \text{ for } x<p \quad \text{or} \quad y_c = \frac{m}{(1-p^2)}((1-2p) + 2px - x^2) \text{ for } x>p \quad (3)$$

For this cambered airfoil, the coordinates (x_U, y_U) and (x_L, y_L), of respectively the upper and lower airfoil surface, become:

$$x_U = x - y_t \sin\theta \ , \quad y_U = y_c + y_t \cos\theta \quad (4)$$

$$x_L = x + y_t \sin\theta \ , \quad y_L = y_c - y_t \cos\theta \quad (5)$$

where

$$\theta = \arctan(\frac{dy_c}{dx}) \quad (6)$$

2.2. To calculate influence coefficients and solve the linear algebraic equations

The linear strength vortex panel method introduced here has the feature that the circulation density on each panel varies linearly from one corner to the other and is continuous across the corner [5]. The sequence of boundary points (end point of each panel) named in the clockwise direction, starting from the trailing edge. The condition that the airfoil be a streamline is met approximately by applying the condition of zero normal velocity components at control point, specified as the mid-point of the panels.

The velocity potential at the ith control point (x_i, y_i) can be described as:

$$\phi(x_i, y_i) = V_\infty(x_i \cos\alpha + y_i \sin\alpha) - \sum_{j=1}^{m} \int_j \frac{\gamma(s_j)}{2\pi} \tan^{-1}\left(\frac{y_i - y_j}{x_i - x_j}\right) ds_j \tag{7}$$

where

$$\gamma(s_j) = \gamma_j + (\gamma_{j+1} - \gamma_j)\frac{s_j}{S_j} \tag{8}$$

in which

V_∞ – velocity of uniform flow, in this article, V_∞ = 1 m/s;

α – angle of attack;

m – vortex panel numbers

(x_j, y_j) – arbitrary point on the jth panel;

S_j – length of jth panel;

s_j – distance measured from the leading edge of the panel;

The no penetration boundary condition requires that the normal velocity of the ith panel at the ith control point be zero, so that

$$\frac{\partial}{\partial n_i}\phi(x_i, y_i) = 0 \quad \text{i=1,2,...,m} \tag{9}$$

Carrying out the involved differentiation and integration, we obtain

$$\sum_{j=1}^{m}(C_{n1_{ij}}\gamma_j' + C_{n2_{ij}}\gamma_{j+1}') = \sin(\theta_i - \alpha) \quad \text{i= 1, 2, ... , m} \tag{10}$$

in which $\gamma' = \gamma / 2\pi V_\infty$ is a dimensionless circulation density, θ_i is the orientation angle between the ith panel and x axis. The coefficients above can be expressed by

$$C_{n1_{ij}} = 0.5DF + CG - C_{n2_{ij}} \tag{11}$$

$$C_{n2_{ij}} = D + 0.5QF / S_j - (AC + DE)G / S_j \tag{12}$$

in which

$$A = -(x_i - X_j)\cos\theta_j - (y_i - Y_j)\sin\theta_j \tag{13}$$

$$B = (x_i - X_j)^2 + (y_j - Y_j)^2 \tag{14}$$

$$C = \sin(\theta_i - \theta_j) \tag{15}$$

$$D = \cos(\theta_i - \theta_j) \tag{16}$$

$$E = (x_i - X_j)\sin\theta_j - (y_i - Y_j)\cos\theta_j \tag{17}$$

$$F = \ln\left(1 + \frac{S_j^2 + 2AS_j}{B}\right) \tag{18}$$

$$G = \tan^{-1}\left(\frac{ES_j}{B + AS_j}\right) \tag{19}$$

$$P = (x_i - X_j)\sin(\theta_i - 2\theta_j) + (y_i - Y_j)\cos(\theta_i - 2\theta_j) \tag{20}$$

$$Q = (x_i - X_j)\cos(\theta_i - 2\theta_j) + (y_i - Y_j)\sin(\theta_i - 2\theta_j) \tag{21}$$

To ensure a smooth flow at the trailing edge, the Kutta condition is applied that,

$$\gamma_1' + \gamma_{m+1}' = 0 \tag{22}$$

Now, there are (m+1) equations and they are sufficient to solve for the (m+1) unknown γ_j' values. We may rewrite the linear equation:

$$\sum_{j=1}^{m+1} A_{n_{ij}} \gamma_j' = (RHS)_i \quad \text{i= 1, 2, …, m+1} \tag{23}$$

in which, for i<m+1:

$$A_{n_{i1}} = C_{n1_{i1}} \tag{24}$$

$$A_{n_{ij}} = C_{n1_{ij}} + C_{n2_{ij-1}} \quad \text{j= 2, 3, …, m} \tag{25}$$

$$A_{ni_{m+1}} = C_{n2_{im}} \tag{26}$$

$$(RHS)_i = \sin(\theta_i - \alpha) \tag{27}$$

for i=m+1:

$$A_{n_{i1}} = A_{n_{im+1}} = 1 \tag{28}$$

$$A_{n_{ij}} = 0 \quad \text{j= 2, 3, …, m} \tag{29}$$

$$(RHS)_i = 0 \tag{30}$$

This system of linear equation is easy to solve by Matlab.

2.3. To predict lift coefficients and pressure distribution of the airfoil

With the known circulation densities, the velocity at each control point can be computed. Since there is only tangential velocity at the control point, the dimensionless velocity defined as $(\partial\phi / \partial t_i)/V_\infty$ can be obtained, which can be given by

$$V_i = \cos(\theta_i - \alpha) + \sum_{j=1}^{m} (C_{t1_{ij}} \gamma_j' + C_{t2_{ij}} \gamma_{j+1}') \quad \text{i= 1, 2, …, m} \tag{31}$$

in which,

$$A_{t_{i1}} = C_{t1_{i1}} \tag{32}$$

$$A_{t_{ij}} = C_{t1_{ij}} + C_{t2_{ij-1}} \quad \text{j= 2, 3, …, m} \tag{33}$$

$$A_{ti_{m+1}} = C_{t2_{im}} \tag{34}$$

After the determination of velocity at control point, the pressure distribution can be computed as

$$C_{pi} = 1 - V_i^2 \tag{35}$$

The lift coefficients can be computed from the pressure distribution by integral below

$$C_l = -\oint C_p \vec{n} \cdot \vec{l} ds \tag{36}$$

3. Results

The results of linear strength vortex panel method for NACA 4412 Airfoil is validated by Xfoil. Figure 1 below shows a C_l vs. angle of attack for vortex panel method (Panel Number N= 160), Xfoil and experimental data. The results from vortex panel method agree reasonably well with those from Xfoil, indicating that the vortex panel code is working well. However, the inviscid solution from panel method becomes increasingly less accurate at higher angles of attack (when $\alpha > 15°$) due to viscous effects. For experimental data, when lift coefficients reach its maximum, $(C_l)_{\max}$, the corresponding angle of attack, α, called the stall angle. After this angle, the lift coefficients from panel method still increase, while the experimental lift coefficients begin to reduce and finally the airfoil stalls.

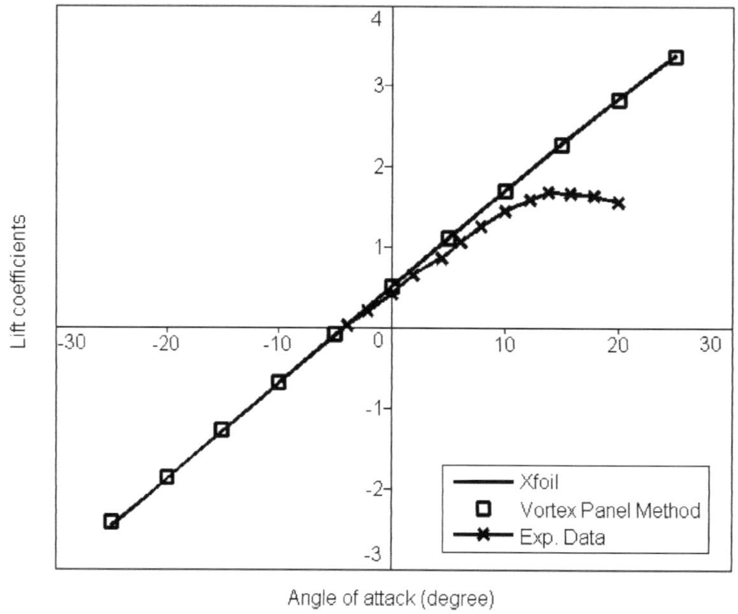

Figure 1. C_l vs. α for NACA 4412 Airfoil

Figure 2 shows the pressure distribution of NACA 4412 Airfoil from linear vortex panel code (Panel Number N= 160). The results are also validated against Xfoil and demonstrate expected shape, which confirms the validity of the code in further step.

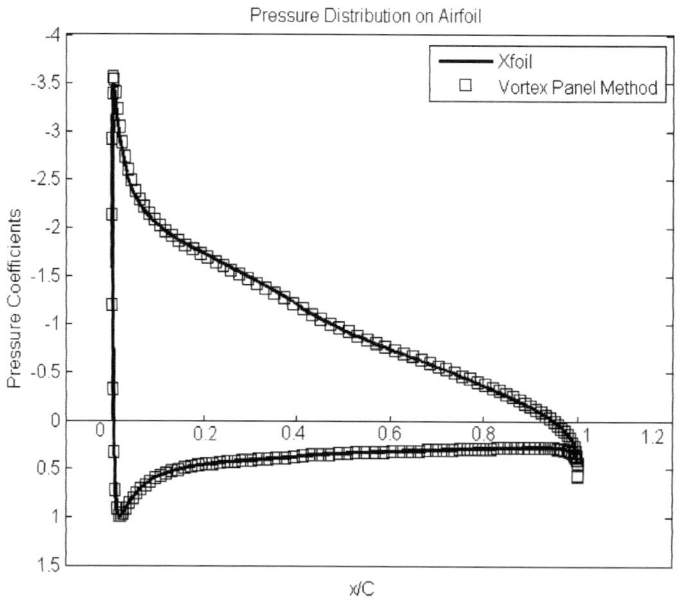

Figure 2. Pressure Distribution for NACA 4412 Airfoil When α =8°

The sensitivity of results to the number of panels is showed below in Figure 3. It is obvious that when panel number N ranges from 10 to 160, C_l is sensitive to N. However, with the increasing panel number (N>160), the results become relatively insensitive to it.

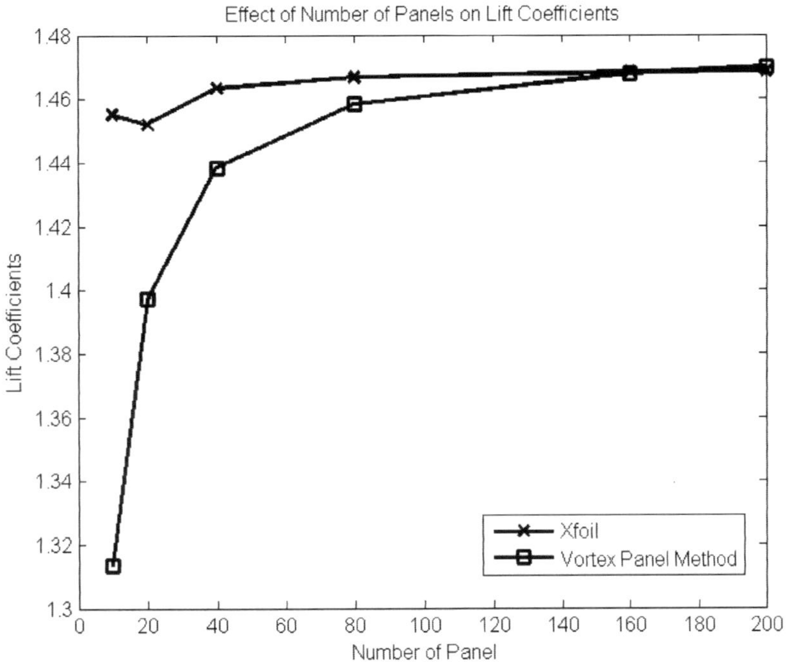

Figure 3. Sensitivity of Results to the Number of Panels When $\alpha = 8^{\circ}$

4. Conclusions

A numerical model for linear strength vortex panel method is developed. By analyzing NACA 4412 Airfoil, the lift coefficients, pressure distribution and sensitivity of results to panel number are studied. The results are validated against Xfoil program, which shows the validity of the vortex panel code.

The analysis results illustrate that panel method cannot account for viscous effects on lift when angle of attack increasing (typical when $\alpha > 15^{\circ}$), which can result in stall. Besides, sensitivity of results to the number of panels N is obvious when N ranging from 10 to 160. When N>160, the results become relatively stable. Thus, when panel method are used to compute complex models, such as large three dimensional model, the number of panels can be chosen appropriately to balance the precision of results and computing time.

Reference

[1] Houghton, E. L. (2013). *Aerodynamics for engineering students*. 6th ed. Waltham, MA: Butterworth-Heinemann.

[2] Flandro, G. A., McMahon, H. M. (2012). *Basic aerodynamics: incompressible flow*. Cambridge: Cambridge University Press.

[3] Katz, J., Plotkin, A., *Low-Speed Aerodynamics*, 2nd ed., Cambridge University Press, 2001.

[4] R. L. Fearn, "Airfoil Aerodynamics Using Panel Methods," *The Mathematica Journal*, Vol. 10, pp. 725-738, 2008.

[5] Kuethe, A. M., Chow, C. (1998). *Foundations of aerodynamics: bases of aerodynamic design*. 5th ed. New York: J. Wiley.

Active Robust Control of Elastic Blade Element Containing Magnetorheological Fluid

Selim Sivrioglu[1] and Fevzi Cakmak Bolat[1]

[1] Department of Mechanical Engineering, Gebze Technical University
Gebze 41400-TURKEY

s.selim@gtu.edu.tr (corresponding author), fcakmakbolat@gmail.com

Abstract. This research study proposes a new active control structure to suppress vibrations of a small-scale wind turbine blade filled with magnetorheological (MR) fluid and actuated by an electromagnet. The aluminum blade structure is manufactured using the airfoil with SH3055 code number which is designed for use on small wind turbines. An interaction model between MR fluid and the electromagnetic actuator is derived. A norm based multi-objective H_2 / H_∞ controller is designed using the model of the elastic blade element. The H_2 / H_∞ controller is experimentally realized under the impact and steady state aerodynamic load conditions. The results of experiments show that the MR fluid is effective for suppressing vibrations of the blade structure.

1. Introduction

The engineering flexible structures like beams, plates and shells show modal dynamical behavior with external disturbances. It is a desired characteristic for such engineering structures to have adaptive behavior for changing external conditions. The control of adaptive structures also affects the dynamics of the flexible systems. Active vibration control approaches improve the performance in terms of reduction in vibration amplitude of the structures compared with passive systems.

In recent years, magnetorheological(MR) materials have been attracted much attention for use in engineering systems. MR fluids contain ferromagnetic particles that can change its rheological properties under application of an external magnetic field. The MR fluid has distinct properties for engineering applications and can be used in vibration control studies [1-3]. In literature, many research works have investigated the behavior of sandwich structures containing MR layer under magnetic field and active vibration control [4-6]. In this study, the ferromagnetic feature of the MR fluid is employed for an electromagnet actuator. The MR fluid is filled into a flexible blade element and attracted by the electromagnet actuator to suppress the vibration of the flexible structure.

2. Elastic blade structure

The blade structure is designed based on the airfoil with SH3055 code number and characteristics and wind tunnel aerodynamic tests for this airfoil is reported in reference [7]. The blade is fixed in its one end with a clamp using a connecting piece fitted with the profile. The length of the blade after clamping is 800 mm and the width of the blade is constant in lengthwise. The cross section and main dimensions of the blade are shown in Figure 1(a). The empty profile of the blade is filled with MR 122EG fluid and both ends are closed with leak proof stick. The amount of MR fluid filled inside the blade profile is 87 ml. Since the structure of the blade is quite complicated, an equivalent beam with MR fluid is designed for modeling purposes. While the length of the equivalent beam is taken as the length of the original blade element, the width and thickness of the equivalent beam is selected so that the geometrical moment of inertia of the beam is equal to the geometrical moment of inertia of the original blade. In the designed equivalent beam, the inner volume contains 107.52 ml MR fluid.

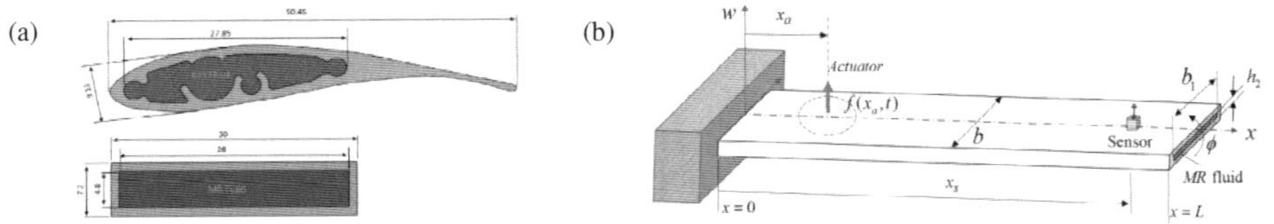

Figure 1. (a) Cross section and dimensions of the blade profile (b) the equivalent beam model

2.1 Equivalent Beam Model

The determination of the eigenfrequencies and the corresponding eigenfunctions is necassary to analyze vibration problem in a distributed system. A modal analysis is realized for the considered elastic blade system. The equivalent beam with MR material considered in this study is schematically illustrated in Figure 1(b). Hamilton principle can be used to derive the equations of motion of the elastic blade element. The equations of motion are obtained as follow

$$\rho\frac{\partial^2 w(x)}{\partial t^2}+2E_f I_f\frac{\partial^4 w(x)}{\partial x^4}-G^* b_1 h_2\left(\frac{\partial^2 w(x)}{\partial x^2}-\frac{\partial \phi}{\partial x}\right)=f(x,t) \tag{1}$$

$$J\frac{\partial^2 \phi}{\partial t^2}-\frac{bh_1 E_f (h_1+h_2)^2}{2}\frac{\partial^2 \phi}{\partial x^2}-G^* bh_2\left(\frac{\partial w}{\partial x}-\phi\right)=0 \tag{2}$$

where $w(x)$ is transverse displacement of the blade and $\phi(x)$ is the cross-sectional rotation. For the cantilever beam, the eigenfunctions are described by

$$\begin{aligned} W_n(x) &= A_1\cosh\lambda_n x+A_2\sinh\lambda_n x+A_3\cos\lambda_n x+A_4\sin\lambda_n x \\ \Phi_n(x) &= C_n(A_1\sinh\lambda_n x+A_2\cosh\lambda_n x+A_3\sin\lambda_n x+A_4\cos\lambda_n x) \end{aligned} \tag{3}$$

where C_n and λ_n are defined as

$$C_n=\frac{2G^* h_2\lambda_n}{E_f h_1(h_1+h_2)^2\lambda_n{}^2+2G^* h_2},\quad \lambda_n=\left(\frac{2n-1}{2}\pi+e_n\right)\frac{1}{L} \tag{4}$$

The vibration response of the blade element is calculated by

$$w(x,t)=\sum_{n=1}^{\infty}W_n(x)\,e^{j\omega_n t},\quad \phi(x,t)=\sum_{n=1}^{\infty}\Phi_n(x)\,e^{j\omega_n t} \tag{5}$$

Substituting equations (5) into (1), the following equation is obtained as

$$\sum_{n=1}^{\infty}\left[-\rho\omega_n^2 W_n(x)+2E_f I_f\frac{d^4 W_n(x)}{dx^4}-G^* bh_2\left(\frac{d^2 W_n(x)}{dx^2}-\frac{d\Phi_n(x)}{dx}\right)\right]=0 \tag{6}$$

In this study, the blade structure contains MR fluid and does not fit this condition but the effect of the MR fluid is negligible due to low stiffness effect of fluid. The natural frequencies of the beam containing MR fluid are calculated as follows

$$\omega_n=\sqrt{\frac{2E_f I_f\lambda_n{}^4-G^* b_1 h_2(\lambda_n C_n-\lambda_n{}^2)}{\mu}} \tag{7}$$

2.2 State Space Model

For each vibration mode of the cantilever blade structure, the separated equation of motion is given by

$$\ddot{x}_n(t) + 2\varsigma\omega_n\dot{x}_n(t) + \omega_n^2 x_n(t) = f(t)\psi_n(x_a) \tag{8}$$

where ω_n is the mode natural frequency, ς is the damping coefficient and $\psi_n(\cdot)$ is the mode shape function. The state space equation for each modal behavior is obtained as follows

$$\dot{\bar{x}}_n(t) = A_n\bar{x}_n(t) + B_n u(t) + D_{wn} d(t) \tag{9}$$

where $\bar{x}_n(t)$ is the state vector, A_n is the system matrix, B_n is the control input matrix and $u(t)$ is the control input. Also, $d(t)$ shows the disturbance input. The structure of the state vector and matrices are as follows

$$\bar{x}_n = \begin{bmatrix} x_n(t) \\ \dot{x}_n(t) \end{bmatrix}, A_n = \begin{bmatrix} 0 & 1 \\ -\omega_n^2 & -2\varsigma\omega_n \end{bmatrix}, B_n = \begin{bmatrix} 0 \\ \psi_n(x_a) \end{bmatrix}, D_{wn} = \begin{bmatrix} 0 \\ \psi_n(x_d) \end{bmatrix} \tag{10}$$

A reduced order state space model for the control design study can be obtained by considering the first two modes. The reduced order state space equation is written as

$$\begin{aligned} \dot{x}_r(t) &= A_r x_r(t) + B_r u(t) + D_{wr} d(t) \\ y_r(t) &= C_r x_r(t) \end{aligned} \tag{11}$$

2.3 Force characterization

The interaction between the blade element with MR fluid and electromagnet is illustrated in Figure 2. The electromagnet is positioned over the MR blade element with an air gap (Figure 2(a)). When the current is applied to the electromagnet and a magnetic field is created, the iron particles in the MR fluid become ordered and an attractive force on the blade is generated (Figure 2(b)). At this stage, the blade stands at the nominal position and still not moving. When the current increases the attractive force also increases and the blade moves to the actuator side as shown in (Figure 2(c)). The nonlinear magnetic force is derived as

$$f(t) = k\frac{(i_0 + i_c(t))^2}{\left[w_0 - w(t)\right]^2} \tag{12}$$

where $k = 1/4\mu_0 N^2 A$. The electromagnetic force is linearized around (i_0, w_0) values as follows.

$$f(t) = K_w w + K_i i_c \tag{13}$$

where $K_w = 2k(i_0^2 / w_0^3)$ and $K_i = 2k(i_0 / w_0^2)$.

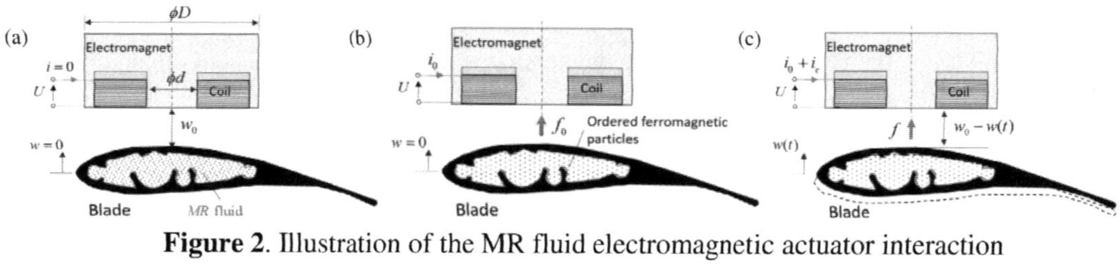

Figure 2. Illustration of the MR fluid electromagnetic actuator interaction

It is important to show how much electromagnetic force is transmitted to the MR blade. To understand the force variation at the MR blade side a load cell is installed under the blade as shown in Figure 3(a). The variations of the experimental forces are shown in Figure 3(b). In these experiments, the air gap between the electromagnet and MR blade is set to 1 mm. At large coil currents, the loss is increasing as seen in Figure 3(b).

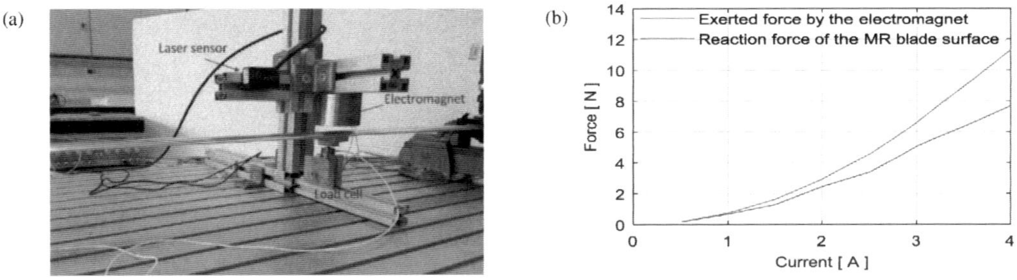

Figure 3. (a) Experimental setup (b)Variation of the electromagnetic force and the response force

3. Multi-objective H_2/H_∞ control

In norm based robust control, H_2 and H_∞ norm performances are two important specifications. While H_∞ performance is convenient to enforce robustness against model uncertainty, H_2 performance is useful to handle stochastic aspects such as measurement noise and control cost [8-9]. $T_\infty(s)$ and $T_2(s)$ are the closed-loop transfer functions from w to z_∞ and z_2, respectively.

The control design block structure is shown in Figure 4(a). In this block, P(s) is the reduced order plant model and K(s) is the controller that will be designed. In the blade system, the input n shows the aerodynamic disturbances and sensor noise. Also, the input d represents the disturbance caused by unstructured uncertainty or unmodeled high frequency dynamics in the system. The controlled variables z_∞ and z_2 are outputs of the transfer functions obtained for each external input. Consider the points p_1, p_2 and p_3 in Figure 4(a). The transfer functions from n to p_1, p_2 and p_3 are obtained as

$$G_{p_1 n}(s) = -\frac{K(s)\varepsilon}{I + P(s)K(s)} = -\varepsilon K(s)S(s), \quad G_{p_3 n}(s) = \frac{\varepsilon}{I + P(s)K(s)} = \varepsilon S(s), G_{p_2 n}(s) = G_{p_1 n}(s) \tag{14}$$

where $S(s) = (I + P(s)K(s))^{-1}$ is the sensitivity transfer function. Moreover, the transfer functions from d to the points p_1, p_2 and p_3 are given as

$$G_{p_1 d}(s) = -\frac{P(s)K(s)}{I + P(s)K(s)} = -T(s), \quad G_{p_3 d}(s) = \frac{P(s)}{I + P(s)K(s)} = T_c(s), G_{p_2 d}(s) = G_{p_1 d}(s) \tag{15}$$

Here $T(s)$ is the complimentary sensitivity transfer function and $T_c(s)$ is the settling function. In a control system design, shaping with $T(s)$ transfer function is preferable for noise attenuation and tracking. Also, $T(s)$ transfer function is important for robust stability with respect to multiplicative uncertainties at the system output. In control block structure, the control cost is adjusted both H_2 and H_∞ norm objectives. The objective of the control design configuration is to minimize

$$\left\| \begin{bmatrix} W_e(s)\varepsilon S(s) \\ W_1(s)\varepsilon K(s)S(s) \\ W_e(s)T_c(s) \\ W_1(s)T(s) \end{bmatrix} \right\|_\infty < \gamma_0 \quad \text{and} \quad \left\| \begin{bmatrix} W_2(s)\varepsilon K(s)S(s) \\ W_2(s)T(s) \end{bmatrix} \right\|_2 < \nu_0 \tag{16}$$

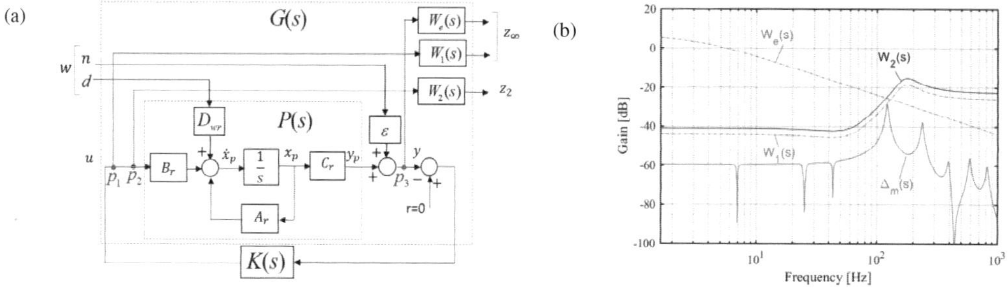

Figure 4. (a) control design structure (b) weighting filters and uncertainty

4. Experimental system

The photo and layout of the experimental system setup is shown in Figure 5. The elastic blade test system is studied for the aims of vibration suppression. In the experimental system, a current driver is used for the electromagnetic actuator. Vibration analysis of the blade is performed with a Bruel&Kjaer 3053 device. The designed controller is realized using dSpace 1104 control card. The controller is discretized and compiled in the state space form using a Matlab/Simulink file and installed on dSpace control card.

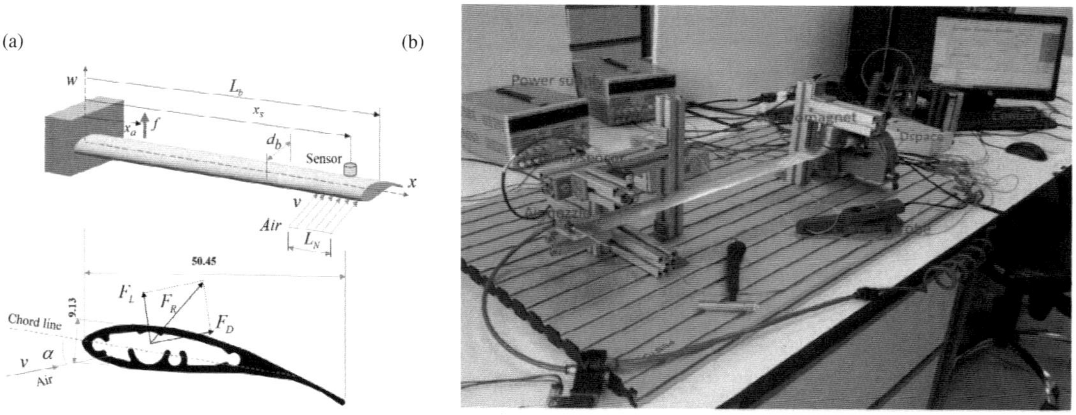

Figure 5. Experimental system setup (a) layout of the blade (b) test system

4.1. Experimental Results

The closed loop frequency responses of the beam obtained in experiments with H_2 / H_∞ controller is shown in Figure 6. As seen in Figure 6(a) responses, the uncontrolled modes are not excited by the controller.

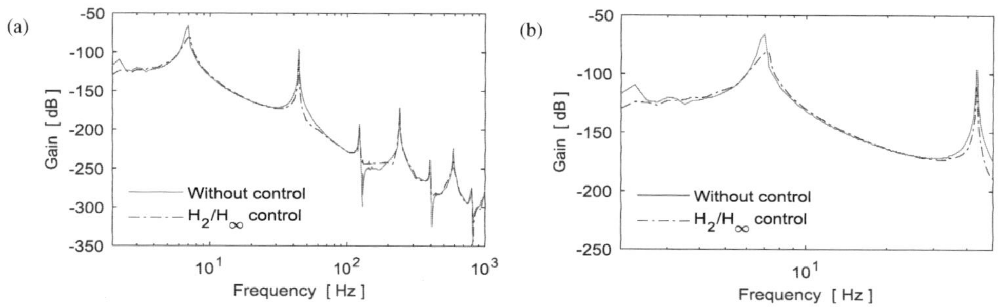

Figure 6. Experimental frequency responses of the closed loop (a) all modes (b) first two modes

The designed multi-objective controller is tested in different conditions under the steady state aerodynamic load. Experimental time history responses of the closed loop system for a continuous control case from a starting time is shown in Figure 7(a). Moreover, the repeated controlled and uncontrolled tests are realized to understand the response characteristics of the controllers as given in Figure 7(b). A robustness test is performed in the case of parameter variation by attaching an extra mass on the surface of the blade. The results are shown in Figure 8 when an additional mass of 12% of the blade mass is added. The control system is quite robust against parameter uncertainty and vibration attenuation with additional mass.

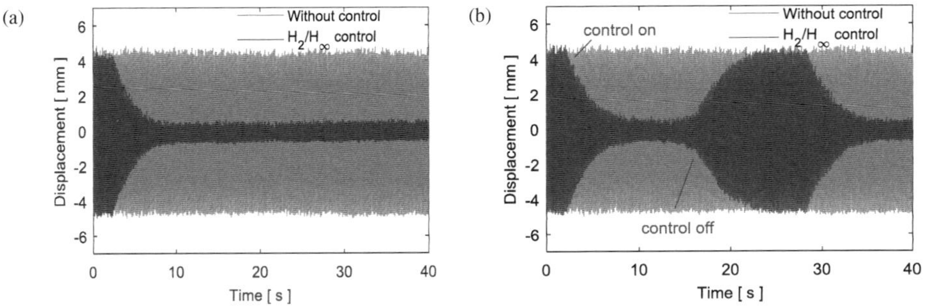

Figure 7. Experimental results of the closed loop system under steady state aerodynamic load

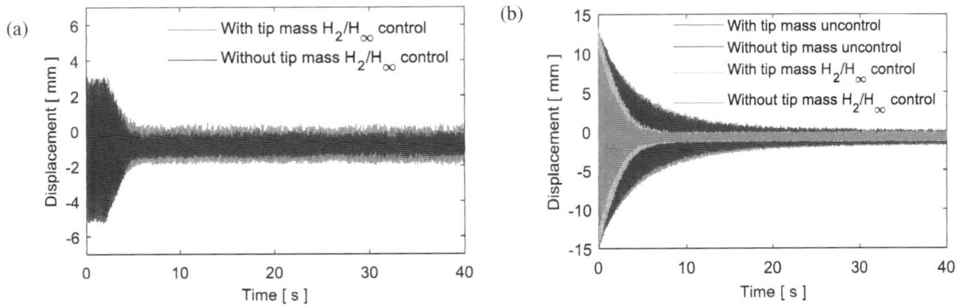

Figure 8. Robustness test results for additional mass, (a) Steady state case, (b) Reference point case

5. Conclusions

In this research study, vibration of a small-scale wind turbine blade is suppressed using a MR blade-electromagnetic actuator under the effect of steady state aerodynamic disturbance. A force based interaction model between MR blade and electromagnet is derived and some characterization works are presented. An H_2 / H_∞ controller is designed to attenuate the vibration of the blade structure. Some experiments are realized to show the effectiveness of the proposed MR blade electromagnetic actuator for the transient and steady state aerodynamic loads. The experimental frequency response and the time history of the closed loop system showed significant vibration reduction in the blade element.

References

[1] Sun, Q., Zhou, J. X., & Zhang, L. (2003). An adaptive beam model and dynamic characteristics of magnetorheological materials. *Journal of Sound and Vibration*, *261*(3), 465-481.

[2] Yalcintas, M., & Dai, H. (1999). Magnetorheological and electrorheological materials in adaptive structures and their performance comparison. *Smart Materials and Structures*, *8*(5), 560.

[3] Weiss, K. D., Carlson, J. D., Nixon, D. A., 1994. Viscoelastic properties of magneto-and electro-rheological fluids, *Journal of Intelligent Material Systems and Structures* 5, p. 772-775.

[4] Niu, H., Zhang, Y., Zhang, X., & Xie, S. (2010). Active vibration control of plates using electro-magnetic constrained layer damping. *International Journal of Applied Electromagnetics and Mechanics*, *33*(1, 2), 831-837.

[5] Valevate, A. V. (2004). Semi-active Vibration Control of a Beam Using Embedded Magneto-rheological Fluids (Doctoral dissertation, Wright State University).

[6] Rajamohan, V., Sedaghati, R., & Rakheja, S. (2011). Optimal vibration control of beams with total and partial MR-fluid treatments. *Smart materials and structures*, *20*(11), 115016.

[7] Selig, M. S., and McGranahan, B. D. (2004) Wind tunnel aerodynamic tests of six airfoils for use on small wind turbines. *Journal of Solar Energy Engineering (Transactions of the ASME)*, *126*(4): 986-1001.

[8] Sivrioglu, S., Nonami, K. (1997) Active Vibration Control by Means of LMI-Based Mixed $H_2/H\infty$ State Feedback Control, *JSME International Journal Series* C 40(2): 239-244.

[9] Madoyan A. (2009) Design and Comparison of Mixed H_2 / H_∞ controller for AMB System, Master Thesis.

IOP Conf. Series: Materials Science and Engineering **326** (2018) 012018 doi:10.1088/1757-899X/326/1/012018

Flame Dynamics inside Rectangular Meso scale Channels

Muzammil Anwar, Abdul Naseer Mohammad, Akram Mohammad and Khalid A Juhany

Department of Aeronautical Engineering, King Abdulaziz University, Jeddah, Saudi Arabia

akramnmania@gmail.com

Abstract. The present work is focused on the experimental study of flame dynamics in preheated meso-scale straight channels of various aspect ratios (2, 5, 12 and 15) and inlet dimensions.Premixed methane-air mixture were used for the reported experiments. To maintain a positive wall temperature gradient inside the channel, the lower part of the rectangular channels were heated at a constant temperature using an external electric heater. Laminar premixed flames were stabilized inside these channels. Various flame propagation modes such as concave, planar, and convex flames with respect to unburned mixture. Concave flames lead to flashback whereas convex flames lead to blowout. Increase in aspect ratio and decrease of flow velocity leads to flame flashback.

1. Introduction

The advancement of combustion systems such as gas turbine combustors, spark ignition engines etc. motivated a large part of the combustion community over the last decade to study the flame propagation using meso scale channels [1]. This inspired the researchers globally to get deeper understanding of the combustion processes such as reactivity, diffusivity, and exothermicity of various fuel-air mixtures. Through the gaps of the order of quenching distance, flame propagation is desirable to control in these micro combustors [2]. The flame front could not propagate in such narrow gaps because of the heat losses occurred from flame front to walls due to larger volume to surface area ratio, which increases proportionally as the combustor size decreases and can make the flame unstable which eventually results in flame quenching. Stable flame is required in the combustor to work efficiently .Therefore several methods have been proposed by the researchers to overcome this difficulty.

Over the years, numerous flame propagation studies have been carried out to upgrade the performance of meso-scale combustors. Mysterious issues in premixed flame propagation in micro or meso scale combustor has not yet been completely unlocked. Some of the critical issues associated to flame propagation in very small channels are thermal diffusive and hydrodynamic inconsistency of a propagating flame [3], formation of boundary layers, flame-wall interaction, hydrodynamic flame-wall interaction induced by wall friction, radical quenching [4-8].To reveal the phenomena of flame propagation characteristics in meso scale channels, Chao et al. [9] have experimented flame stabilization using cylindrical quartz tubes with varying diameters. Similar study was carried out by Maruta et al. [4] on characteristics of methane-air premixed flames propagation in cylindrical quartz tube with a wall temperature gradient along the fluid flow direction. They achieved stable flames in very low and high flow rate regions .Other dynamic properties such as flame with repetitive extinction and ignition (FREI) and fluctuating flames were reported at moderate flow rates. Several propagation regimes such as flashback, static flame front, flickering and blow out were reported by Evans and Kyritsis [10] in their study of non-adiabatic ducts on meso-scale for both methane and propane. Numerical simulations of flames in micro scale channels were also studied by Pizza [5] with lean premixed mixture of hydrogen/air with a constant wall temperature and height varying from 0.3 to 1.0mm.They had also concluded the existence of flames with repetitive extinction and ignition (FREI) for low inflow velocity conditions. Further increase in the flow velocity resulted in the formation of a stable asymmetric flame. A similar study was carried out by Pizza et al. [6] showed some new flame propagating modes like open and closed both stable and unstable, symmetric and asymmetric for meso scale channels of height ranging between 2.0 and 7.0mm.Characterstics of flame propagation were investigated by Khandelwal and Kumar [11] using rectangular diverging channels. This work shows different propagation modes of a flame in straight rectangular channels of 2, 5, 12, and 15 aspect ratios. This study is essential from application viewpoint because some flame propagation modes like partially stable flames can lead to poor combustion resulting in loss of combustion efficiency. . The main goal of this study is to minimize formation of unstable flames in meso-scale combustors and to conserve maximum thermal energy at the flame area. Therefore a systematic detailed study on the dynamics of flames in straight rectangular channels with different aspect ratios is presented.

2. Experimental method

2.1. Experimental setup

The schematic diagram of the experimental set-up with detailed dimensions of a typical 15 aspect ratio channel is shown in figure 1. A premixed Ch4–air mixture is supplied through the inlet of the channel at 300 K. The methane-air flow rates can be controlled using Aalborg mass flow controllers which are connected to the PC. Rectangular straight mesoscale channels of 2, 5, 12, 15 aspect ratios were chosen. The inlet height was maintained at 2mm and the width was varied . Quartz is chosen because of its low thermal conductivity and expansion, high heat capacity and transparency of the walls to help visualize the flame front location in the channel. The investigations reported in this paper has been carried out using methane-air mixture. The channels were placed exactly 10 mm above the heater and the overlap distance was also kept approximately 10 mm from the exit of the channel. With the help of a level indicator the channel was kept parallel to the burner within an accuracy of ±0.1. The quartz channel was heated constantly throughout the experimental investigations to exclude the effect of varying temperature distribution with heating rate on the observed flame patterns. The rate of heat release through the combustion of methane–air mixture is relatively very small (60 W) as compared to the heat release rate of sintered metal burner (800 W) which ensures thermally thick walls and avoids thermal coupling between solid and gas phase.

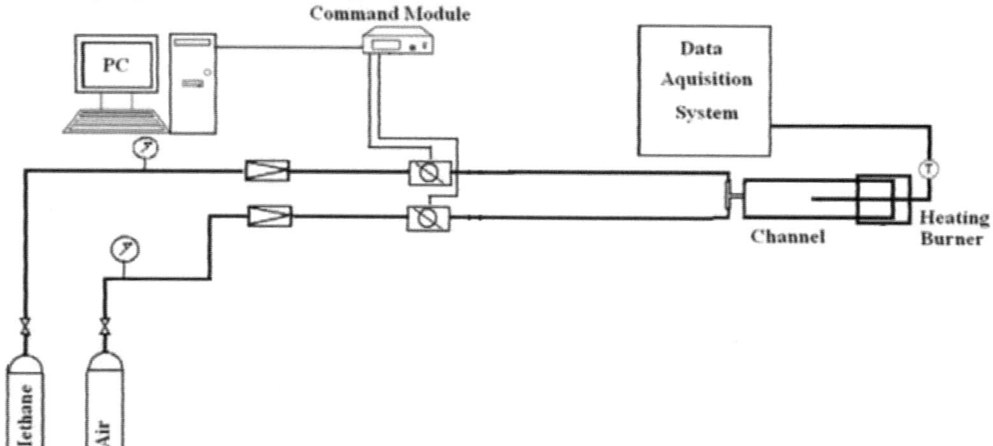

Figure 1: Schematic of experimental setup

3. Results and Discussion

In the present experiments, various flame propagation modes are observed. These propagation modes are classified into three different categories based on their appearance as planar flames, concave flames and convex flames

3.1 Planar flames

Flame propagation modes at different flow velocities in low-aspect-ratio channels (2 and 5) were observed to be flat/planar. Flat shaped flame propagation mode were spotted for a range of mixture equivalence ratios Φ, varying from 0.8 to 1.1 and inlet mixture velocity depending upon the external heating rates supplied to preheat the channel walls.

Photographs of planar flames observed inside low-aspect-ratio channel for various equivalence ratios and flow velocities are shown in figure 2. The mixture is ignited at the exit of the channel and the flame front moves inside. The flame front then becomes planar and completely stable at a location where the flow velocity matches the burning velocity of the mixture at that temperature. The location of flame depends on the mixture inlet velocity and external heating rates.

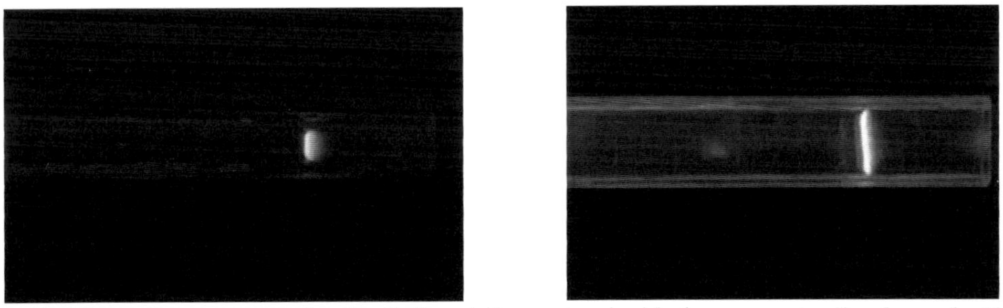

Figure 2: Planar flames spotted with different flow rates inside low-aspect-ratio channels (AR 2 and 5)

3.2 Concave and Convex flames:

The flame front for certain conditions of the inlet velocity and mixture equivalence ratios are of concave and convex shape, termed concave and convex flames. The photographs of these concave and convex flames in are shown in figures 3a and 3b for Φ=1.0 and U_{inlet}=0.55m/s U_{inlet}=0.6m/s in 12° and 15° aspect ratio channel. The concave flames were observed in very high-aspect-ratio channels only (aspect ratio>10) at moderate flow rates, whereas Convex flames were observed in all channels. In high-aspect-ratio channels, convex stretched flames were observed at high inlet velocities

Figure 3a: Photographs showing concave flames in AR 12 and 15 channels at U_{inlet}=0.55 m/s

Figure 3b: Photographs showing convex flames in AR 12 and 15 channels at U_{inlet}=0.6 m/s

3.3 Effect of mixture equivalence ratio Φ on flames

The mixture equivalence ratio effect on the position of the flame in various aspect ratio channels is shown in figure 4. It is observed that the flames were stabilized near the exit of the channel for rich and very lean mixtures. For mixtures within the range of (Φ 0.9~1.0) the flames stabilized inside the channels away from the exit and closer to the inlet. It is also observed that the flames at Φ=1.1 propagated much faster than the flames of mixtures of other equivalence ratios. This is because of the low burning velocities of these mixtures in comparison to mixtures near stoichiometric conditions.

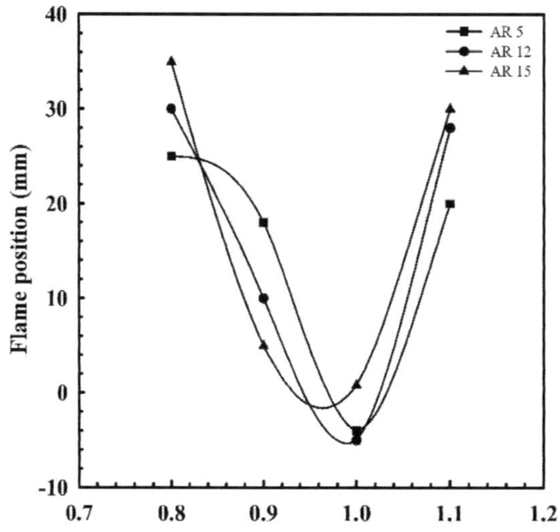

Figure 4: Effect of equivalence ratio Φ on flame position in various channels

3.4 Effects of flow inlet velocity on flames

Effects of flow inlet velocity on the stabilization of flames inside the channels were investigated in the axial direction for 5AR channel with varying mixture equivalence Φ ratios in figure 5. The flame gets stabilized at a place where the inlet flow velocity matches the burning velocity of the mixture. Changes occurred in the location of flame stabilization depends on factors such as temperature gradient and heat loss from flame to solid walls effects the location of flame stabilization.

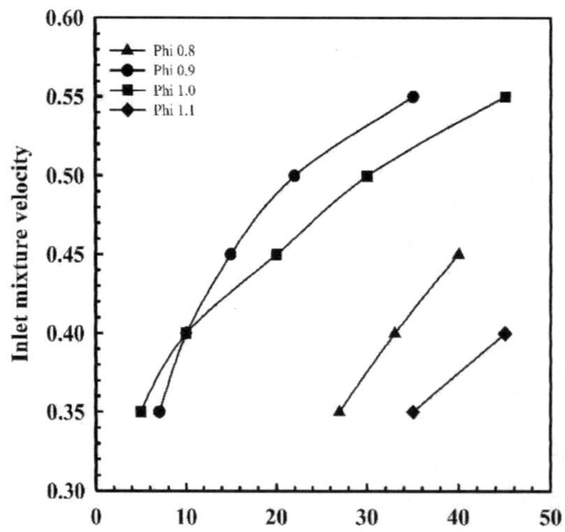

Figure 5: Effect of flow velocity on flame position with varying equivalence ratios for 5AR channel

IOP Conf. Series: Materials Science and Engineering **326** (2018) 012018 doi:10.1088/1757-899X/326/1/012018

4. Conclusion

In this article, flame dynamics has been studied experimentally using preheated mesoscale straight channels of various aspect ratios. Premixed methane-air mixtures were used to carry out the investigations for a range of operating conditions by varying both equivalence ratios and inlet flow rates. The lower area of the quartz channels were heated at a constant temperature with an external heating source burner which resulted in formation of linear temperature gradient along the axial direction inside the channels. Three types of flame propagation modes were observed during the experiments planar flames, concave flames and convex flames. Stretch free flames were observed only for low aspect ratio channels at moderate flow rates. These flames were observed to be symmetric or asymmetric to the x axis. Convex flames were observed in high aspect ratio channels (AR 12 and 15) with high inflow rates more than 0.5 m/s. The investigations showed the existence of concave flames in high aspect ratio channels at flow rates less than 0.5m/s for all equivalence ratios. Flames with repetitive extinction and ignition (FREI) was observed at very low mass flow rates.

References

[1] Fernandez-Pello, A.C., *Micropower generation using combustion: issues and approaches.* Proceedings of the combustion institute, 2002. 29(1): p. 883-899.

[2] Akram, M. and S. Kumar, Experimental studies on dynamics of methane–air premixed flame in meso-scale diverging channels. Combustion and flame, 2011. 158(5): p. 915-924.

[3] Bychkov, V. and M.A. Liberman, *Dynamics and stability of premixed flames.* Physics reports, 2000. 325(4): p. 115-237.

[4] Maruta, K., et al., *Characteristics of combustion in a narrow channel with a temperature gradient.* Proceedings of the Combustion Institute, 2005. 30(2): p. 2429-2436.

[5] Pizza, G., et al., *Dynamics of premixed hydrogen/air flames in microchannels.* Combustion and Flame, 2008. 152(3): p. 433-450.

[6] Pizza, G., et al., *Dynamics of premixed hydrogen/air flames in mesoscale channels.* Combustion and Flame, 2008. 155(1): p. 2-20.

[7] Kurdyumov, V.N., et al., Dynamics of premixed flames in a narrow channel with a step-wise wall temperature. Combustion and Flame, 2009. 156(11): p. 2190-2200.

[8] Ronney, P.D., *Analysis of non-adiabatic heat-recirculating combustors.* Combustion and Flame, 2003. 135(4): p. 421-439.

[9] Chao, C.Y.H., et al., *Analytical and experimental study of premixed methane–air flame propagation in narrow channels.* International journal of heat and mass transfer, 2007. 50(7): p. 1302-1313.

[10] Evans, C.J. and D.C. Kyritsis, *Operational regimes of rich methane and propane/oxygen flames in mesoscale non-adiabatic ducts.* Proceedings of the Combustion Institute, 2009. 32(2): p. 3107-3114.

[11] Khandelwal, B. and S. Kumar, Experimental investigations on flame stabilization behavior in a diverging micro channel with premixed methane–air mixtures. Applied Thermal Engineering, 2010. 30(17): p. 2718-2723.